Designing Microservices Platforms with NATS

A modern approach to designing and implementing scalable microservices platforms with NATS messaging

Chanaka Fernando

BIRMINGHAM—MUMBAI

Designing Microservices Platforms with NATS

Copyright © 2021 Packt Publishing

Group Product Manager: Aaron Lazar

Publishing Product Manager: Harshal Gundetty

Senior Editor: Ruvika Rao

Content Development Editor: Vaishali Ramkumar

Technical Editor: Maran Fernandes

Copy Editor: Safis Editing

Project Coordinator: Deeksha Thakkar

Proofreader: Safis Editing

Indexer: Sejal Dsilva

Production Designer: Roshan Kawale

First published: October 2021

Production reference: 1141021

Published by Packt Publishing Ltd.

Livery Place

35 Livery Street

Birmingham

B3 2PB, UK.

ISBN 978-1-80107-221-2

www.packt.com

To my mother, Seelawathi De Silva, and my father, Harry Sebastian, for their sacrifices and for exemplifying the power of determination. To my wife, Subhashini, for being my loving partner throughout our joint life journey, and to my little daughter, Saneli, for bringing me happiness and joy.

– Chanaka Fernando

Contributors

About the author

Chanaka Fernando is a solution architect with 12+ years of experience in designing, implementing, and supporting enterprise-scale software solutions for customers across various industries including finance, education, healthcare, and telecommunications. He has contributed to the open source community with his work (design, implementation, and support) as the product lead of the WSO2 ESB, one of the founding members of the "Ballerina: cloud-native programming language" project, and his own work on GitHub. He has spoken at several WSO2 conferences and his articles are published on Medium, DZone, and InfoQ.

Chanaka has a bachelor's degree in electronics and telecommunications engineering from the University of Moratuwa.

To my wife, Subhashini, my mother-in-law, Margrett, my father-in-law, Bernard, and my parents. They made sure I had enough time and energy to spend on writing this book. To Harshal, for believing in the book's topic and giving me the opportunity to write for Packt. To Vaishali, Ruvika, Deeksha, and the rest of the Packt team, I thank you all for your tireless efforts. To Justice Nefe and Isuru Udana, both great technical reviewers, I am grateful for your comments and feedback.

About the reviewers

Justice Nefe, is CEO of Borderless HQ, Inc. and has 5+ years' experience as a software engineer with a focus on large-scale systems. Justice has built products and services with tools including Node.js, Golang, Vue.js, Docker, Kubernetes, gRPC, GraphQL, and others; and has designed systems from monoliths to microservices, leveraged on Apache Pulsar/nats.io and NATS Streaming for event-driven microservices. Justice is currently experimenting with Rust and Flutter and is skilled in distributed systems development, enterprise software development, product development, and systems design, working with different teams large and small, as well as teams in different geographical zones all to create products and services that put a smile on the face of customers. Justice has created open source products and internal tools for the different teams they've worked with, and collaborates with key stakeholders to conceptualize and drive new and existing initiatives to market.

Isuru Udana Loku Narangoda is a software architect and associate director at WSO2 with more than 10 years of experience in the enterprise integration space. Isuru is one of the product leads of the WSO2 Enterprise Integrator and API Manager products, and provides technical leadership to the project. Isuru is an open source enthusiast, a committer, and holds the vice-president position of the Apache Synapse open source ESB project. Also, Isuru has participated in the Google Summer of Code program as a student as well as a mentor for several years.

Table of Contents

2

Why Is Messaging Important in the Microservice Architecture?

3

What Is NATS Messaging?

Section 2: Building Microservices with NATS

6

A Practical Example of Microservices with NATS

7

Securing a Microservices Architecture with NATS

8
Observability with NATS in a Microservices Architecture

Section 3: Best Practices and Future Developments

9
How Microservices and NATS Coexist with Integration Platforms

10

Future of the Microservice Architecture and NATS

Other Books You May Enjoy

Index

Preface

The microservices architecture has developed into a mainstream approach to building enterprise-grade applications within the past few years. Many organizations, from large to medium to small start-ups, have started utilizing the microservices architecture to build their applications. With more and more people adopting the microservices approach to build applications, some practical challenges of the architecture have been uncovered. Inter-service communication is one challenge that most microservices teams experience when scaling applications to a larger number of instances.

At first, point-to-point inter-service communication was not working well, and the concept of smart endpoints and dumb pipes was proposed as an alternative approach. Instead of connecting microservices in a point-to-point manner, having a messaging layer to decouple the microservices looked like a better solution.

NATS messaging technology was originally developed as the messaging technology to be used in the Cloud Foundry platform. It was built to act as the always-on dial tone for inter-service communication. Its performance and the simple interface it exposed to interact with clients made it popular within the developer community.

In this book, we discuss how NATS messaging can be used to implement inter-service communication within a microservices architecture. We start with a comprehensive introduction to microservices, messaging, and NATS technology. Then we go through the architectural aspects and provide a reference implementation of an application using the Go programming language. We cover the security and observability aspects of the proposed solution and how that can co-exist in an enterprise platform. At the end of the book, we discuss the latest developments in microservices and NATS messaging and explore how these developments can shape our proposed solution.

Who this book is for

This microservices book is for enterprise software architects and developers who design, implement, and manage complex distributed systems with microservices architecture concepts. Intermediate-level experience of any programming language and software architecture is required to make the most of this book. If you are new to the field of microservices architecture and NATS messaging technology, you can use this book as a learning guide to get into those areas.

What this book covers

Chapter 1, Introduction to the Microservices Architecture, provides a comprehensive introduction to the microservices architecture.

Chapter 2, Why Is Messaging Important in a Microservices Architecture?, discusses different messaging technologies and why microservices architectures require messaging.

Chapter 3, What Is NATS Messaging?, explores the NATS messaging technology by covering the concepts with practical examples.

Chapter 4, How to Use NATS in a Microservices Architecture, discusses the possible ways to use NATS messaging in a microservices context.

Chapter 5, Designing a Microservices Architecture with NATS, provides a reference architecture using a real-world application to build a microservices-based application with NATS.

Chapter 6, A Practical Example of Microservices with NATS, provides a reference implementation of an application using the microservices architecture along with NATS.

Chapter 7, Securing a Microservices Architecture with NATS, discusses the security of the overall microservices architecture, including NATS, with examples on securing NATS servers.

Chapter 8, Observability with NATS in a Microservices Architecture, explores various monitoring and troubleshooting requirements and available technologies with an example implementation.

Chapter 9, How Microservices and NATS Co-exist with Integration Platforms, discusses the aspects related to the integration of microservices-based applications with other enterprise systems.

Chapter 10, Future of the Microservices Architecture and NATS, explores the new developments in the microservices and NATS domains.

To get the most out of this book

This book is written in such a way that you will get the best learning experience by reading the chapters in order. The book includes commands, code examples, and step-by-step instructions as and when necessary. Following these instructions will help immensely in understanding the concepts. The book also provides several exercises so that you can improve your understanding and apply the knowledge to real-world applications. Try to complete the exercises while reading the book.

Software/hardware covered in the book	Operating system requirements
NATS 2.2	Windows, macOS, or Linux
Go 1.16.5	Windows, macOS, or Linux
Prometheus 2.28	Windows, macOS, or Linux
Grafana 8.1.2	Windows, macOS, or Linux
Loki 2.3	Windows, macOS, or Linux
WSO2 API Manager 4.0	Windows, macOS, or Linux
WSO2 Micro Integrator 4.0	Windows, macOS, or Linux
Python 3.6	Windows, macOS, or Linux
Java Runtime Environment (JRE) 11	Windows, macOS, or Linux

In addition to the this software, you need the CFSSL tool to create certificates to try out the examples in Chapter 7, Securing a Microservices Architecture with NATS. This tool can be downloaded from here: `https://github.com/cloudflare/cfssl`.

All the examples in this book were tested using macOS. Most of the examples should work with both Windows and Linux operating systems.

If you are using the digital version of this book, we advise you to type the code yourself or access the code from the book's GitHub repository (a link is available in the next section). Doing so will help you avoid any potential errors related to the copying and pasting of code.

You may benefit from following the author on Twitter (`https://twitter.com/chanakaudaya`), *Medium* (`https://chanakaudaya.medium.com`), *and GitHub* (`https://github.com/chanakaudaya`), *or by adding them as a connection on LinkedIn* (`https://linkedin.com/in/chanakaudaya`).

Download the example code files

You can download the example code files for this book from GitHub at `https://github.com/PacktPublishing/Designing-Microservices-Platforms-with-NATS/`. If there's an update to the code, it will be updated in the GitHub repository.

We also have other code bundles from our rich catalog of books and videos available at `https://github.com/PacktPublishing/`. Check them out!

Code in Action

The Code in Action videos for this book can be viewed at `http://bit.ly/2OQfDum`.

Download the color images

We also provide a PDF file that has color images of the screenshots and diagrams used in this book. You can download it here: `https://static.packt-cdn.com/downloads/9781801072212_ColorImages.pdf`.

Conventions used

There are a number of text conventions used throughout this book.

`Code in text`: Indicates code words in text, database table names, folder names, filenames, file extensions, pathnames, dummy URLs, user input, and Twitter handles. Here is an example: "The request message is published to the `patient.profile` subject and the subscribers are listening on the same subject."

A block of code is set as follows:

```
func main() {
// Initialize Tracing
initTracing()
}
```

Any command-line input or output is written as follows:

```
$ nats-server --config node.conf --log nats.log
```

Bold: Indicates a new term, an important word, or words that you see onscreen. For instance, words in menus or dialog boxes appear in **bold**. Here is an example: "You could observe the **+OK** message coming from the server as a response to the PUB command."

> Tips or important notes
> Appear like this.

Get in touch

Feedback from our readers is always welcome.

General feedback: If you have questions about any aspect of this book, email us at customercare@packtpub.com and mention the book title in the subject of your message.

Errata: Although we have taken every care to ensure the accuracy of our content, mistakes do happen. If you have found a mistake in this book, we would be grateful if you would report this to us. Please visit www.packtpub.com/support/errata and fill in the form.

Piracy: If you come across any illegal copies of our works in any form on the internet, we would be grateful if you would provide us with the location address or website name. Please contact us at copyright@packt.com with a link to the material.

If you are interested in becoming an author: If there is a topic that you have expertise in and you are interested in either writing or contributing to a book, please visit authors.packtpub.com.

Share Your Thoughts

Once you've read *Designing Microservices Platforms with NATS*, we'd love to hear your thoughts! Scan the QR code below to go straight to the Amazon review page for this book and share your feedback.

https://packt.link/r/1-801-07221-3

Your review is important to us and the tech community and will help us make sure we're delivering excellent quality content.

Section 1: The Basics of Microservices Architecture and NATS

This first section provides an understanding of what the microservices architecture is and the benefits of using it for application development. It also discusses different messaging technologies and how these technologies can be utilized to build microservices-based applications. Then, it introduces NATS messaging technology by covering the concepts with practical examples.

This section contains the following chapters:

- *Chapter 1, Introduction to the Microservices Architecture*
- *Chapter 2, Why Is Messaging Important in Microservices Architecture?*
- *Chapter 3, What Is NATS Messaging?*

1
Introduction to the Microservice Architecture

The microservice architecture is an evolutionary approach to building effective, manageable, and scalable distributed systems. The overwhelming popularity of the internet and the smart digital devices that have outnumbered the world's population have made every human being a consumer of digital products and services. Business leaders had to re-evaluate their enterprise IT platforms to make sure that these platforms are ready for the consumer revolution due to the growth of their business. The so-called **digital-native** companies such as Google, Amazon, Netflix, and Uber (to name a few) started building their enterprise platforms to support this revolution. The microservice architecture evolved as a result of the work that was done at these organizations to build scalable, manageable, and available enterprise platforms.

When microservice-based platforms become larger and larger with hundreds or thousands of microservices inside them, these services need to communicate with each other using the point-to-point model before they become too complicated to manage. As a solution to this problem, centralized message broker-based solutions provided a less complex and manageable solution. Organizations that adopted the microservice architecture are still evaluating the best possible approach to solve the problem of communication among services. The so-called model of **smart endpoints and dumb pipes** also suggests using a message broker-based approach for this.

NATS is a messaging framework that acts as the always-on dial tone for distributed systems communication. It supports the traditional pub-sub messaging model, which is supported by most of the message brokers, as well as the request-response style communication model while supporting high message rates. It can be used as the messaging framework for the microservice architecture.

In this book, we will discuss the concepts surrounding the microservice architecture and how we can use the NATS messaging framework to build effective, manageable, and scalable distributed systems.

Distributed computing systems have evolved from the early days of mainframes and large servers, sitting in separate buildings, to serverless computing, where users do not even need to consider the fact that there is a server that is running their software component. It is a journey that continues even today and into the future. From the early scheduled jobs to simple programs written in Assembly to monolithic applications written in C, or from Java to ESB/SOA-based systems to microservices and serverless programs, the evolution continues.

IT professionals have been experimenting with different approaches to solve the complex problem of distributed computing so that it eventually produces the best experience for the consumers. The microservice architecture brings out several benefits to the distributed computing system's design and implementation, which was not feasible before. It became mainstream at a time where most of the surrounding technological advancements, such as containers, cloud computing, and messaging technologies, are also becoming popular. This cohesion of technologies made the microservice architecture even more appealing to solve complex distributed systems-related challenges.

In this chapter, we're going to cover the following main topics:

- The evolution of distributed systems
- What is a microservice architecture?
- Characteristics of the microservice architecture

- Breaking down a monolith into microservices
- Advantages of the microservice architecture

The evolution of distributed systems

The quality of the human mind to ask for more has been the driving force behind many innovations. In the early days of computing, a single mainframe computer executed a set of batch jobs to solve a certain mathematical problem at an academic institution. Then, large business corporations wanted to own these mainframe computers to execute certain tasks that would take a long time to complete if done by humans. With the advancements in electrical and electronics engineering, computers became smaller and instead of having one computer sequentially doing all the tasks, business owners wanted to execute multiple tasks in parallel by using multiple computers. The effects of improved technology on electronic circuits and their reduced size resulted in a reduction in costs, and more and more organizations started using computers.

Instead of getting things done through a single computer, people started using multiple computers to execute certain tasks, and these computers needed to connect to communicate and share the results of their executions to complete the overall task. This is where the term *distributed systems* came into use.

A **distributed system** is a collection of components (applications) located on different networked computers that communicate and coordinate their tasks by passing messages to one another via a network to achieve a common goal.

Distributing a workload (a task at hand) among several computers poses challenges that were not present before. Some of those challenges are as follows:

- Failure handling
- Concurrency
- Security of data
- Standardizing data
- Scalability

Let's discuss these challenges in detail so that the distributed systems that we will be designing in this book can overcome these challenges well.

Failure handling

Communication between two computers flows through a network. This can be a wired network or a wireless network. In either case, the possibility of a failure at any given time is inevitable, regardless of the advancements in the telecommunications industry. As a designer of distributed systems, we should vary the failures and take the necessary measures to handle these failures. A properly designed distributed system must be capable of the following:

- Detecting failures
- Masking failures
- Tolerating failures
- Recovery from failures
- Redundancy

We will discuss handling network and system failures using the preceding techniques in detail in the upcoming chapters.

Concurrency

When multiple computers are operating to complete a task, there can be situations where multiple computers are trying to access certain resources such as databases, file servers, and printers. But these resources may be limited in that they can only be accessed by one consumer (computer) at a given time. In such situations, distributed computer systems can fail and produce unexpected results. Hence, managing the concurrency in a distributed system is a key aspect of designing robust systems. We will be discussing techniques such as messaging (with NATS) that can be used to address this concurrency challenge in upcoming chapters.

Security of data

Distributed systems move data from one computer to another via a communication channel. These communication channels are sometimes vulnerable to various types of attacks by internal and external hackers. Hence, securing data transfers across the network is a key challenge in a distributed system. There are technologies such as **Secure Socket Layer** (**SSL**) that help improve the security of wire-level communication. It is not sufficient in a scenario where systems are exposing business data to external parties (for example, customers or partners). In such scenarios, applications should have security mechanisms to protect malicious users and systems from accessing valuable business data. Several techniques have evolved in the industry to protect application data.

Some of them are as follows:

- **Firewalls and proxies to filter traffic**: Security through network policies and traffic rules.

- **Basic authentication with a username and password**: Protect applications with credentials provided to users in the form of a username and password.

- **Delegated authentication with 2-legged and 3-legged OAuth flow (OAuth2, OIDC)**: Allow applications to access services on behalf of the users using delegated authentication.

- **Two-Factor Authentication (2FA)**: Additional security with two security factors such as username/password and a **one-time password (OTP)**.

- **Certificate-based authentication (system-to-system)**: Securing application-to-application communication without user interaction using certificates.

We will be exploring these topics in detail in the upcoming chapters.

Standardizing data

The software components that are running on different computers may use different data formats and wire-level transport mechanisms to send/receive data to/from other systems. This will become a major challenge when more and more systems are introduced to the platform with different data and transport mechanisms. Hence, adhering to a common standard makes it easier to network different systems without much work. Distributed systems designers and engineers have come up with various standards in the past, such as XML, SOAP, and REST, and those standards have helped a lot in standardizing the interactions among systems. Yet there is a considerable number of essential software systems (such as ERP and CRM) that exchange data with proprietary standards and formats. On such occasions, the distributed system needs to adopt those systems via technologies by using an adapter or an enterprise service bus that can translate the communication on behalf of such systems.

Scalability

Most systems start with one or two computers running a similar number of systems and networking, which is not a difficult task. But eventually, these systems become larger and larger and sometimes grow to hundreds or thousands of computers running a similar or a greater number of different systems.

Hence, it is essential to take the necessary action at the very early stages to address the challenge of scalability. There are various networking topologies available to design the overall communication architecture, as depicted in *Figure 1.1* – Networking topologies. In most cases, architects and developers start with the simplest model of point-to-point and move into a mesh architecture or star (hub) architecture eventually.

The bus topology is another common pattern most of the distributed systems adhered to in the past, and even today, there are a significant number of systems using this architecture.

Distributed systems networking architecture

The software engineers and architects who worked on these initial distributed computing system's designs and implementations have realized that different use cases require different patterns of networking. Therefore, they came up with a set of topologies based on their experiences. These topologies helped the systems engineers to configure the networks efficiently based on the problem at hand. The following diagram depicts some of the most common topologies used in distributed systems:

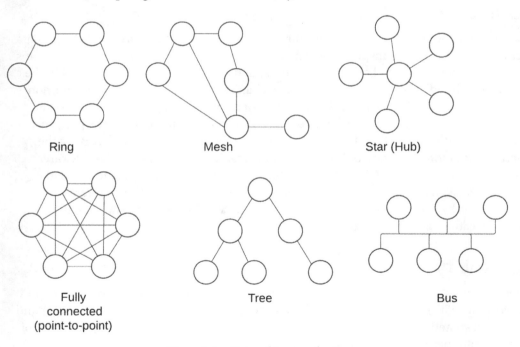

Figure 1.1 – Networking topologies

These topologies helped engineers solve different types of real-world problems with distributed computing. In most cases, engineers and architects started with a couple of applications connected in a point-to-point manner. When the number of applications grows, this becomes a complicated network of point-to-point connections. These models were easy to begin with, yet they were harder to manage when the number of nodes grew beyond a certain limit. In traditional IT organizations, change is something people avoid unless it is critical or near a break-even point. This reserved mindset has made many enterprise IT systems fall into the category of either a mesh or a fully connected topology, both of which are hard to scale and manage. The following diagram shows a real-world example of how complicated an IT system would look like with this sort of topology:

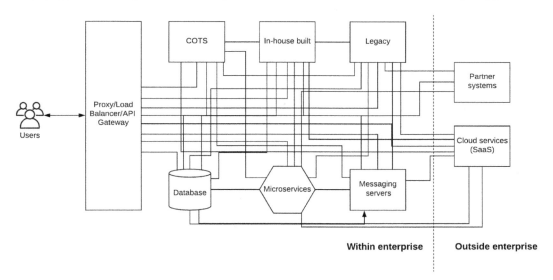

Figure 1.2 – Distributed system with a mesh topology

The preceding diagram depicts an architecture where multiple applications are connected in a mesh topology that eventually became an unmanageable system. There are many such examples in real IT systems where deployments become heavily complicated, with more and more applications being introduced as a part of the business's evolution.

The era of the service-oriented architecture (SOA) and the enterprise service bus (ESB)

The IT professionals who were designing and implementing these systems realized the challenge and tried to find alternative approaches to building complex distributed systems. By doing so, they identified that a bus topology with a clear separation of responsibilities and services can solve this problem. That is where the **service-oriented architecture (SOA)** became popular, along with the centralized **enterprise service bus (ESB)**.

The SOA-based approach helped IT professionals build applications (services) with well-defined interfaces that abstract the internal implementation details so that the consumers of these applications would only need to integrate through the interface. This approach reduced the tight coupling of applications, which eventually ended up in a complex mesh topology with a lot of friction for change.

The SOA-based approach allowed application developers to change their internal implementations more freely, so long as they adhered to the interface definitions. The centralized service bus (ESB) was introduced to network various applications that were present in the enterprise due to various business requirements. The following diagram depicts the enterprise architecture with the bus topology, along with an ESB in the middle acting as the bus layer:

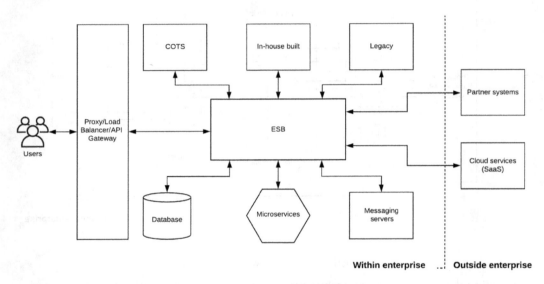

Figure 1.3 – Distributed system with the bus topology using ESB

As depicted in the preceding diagram, this architecture worked well in most use cases, and it allowed the engineers and architects to reduce the complexity of the overall system while onboarding more and more systems that were required for business growth. One challenge with this approach was that more and more complex logic and the load were handled by the centralized ESB component, and it became a central point of failure unless you deployed that with high availability. This was inevitable with this architecture and IT professionals were aware of this challenge.

Scaling for demand

With the introduction of agile development methodologies, container-based deployments, and the popularity of cloud platforms, this ESB-based architecture looked obsolete, and people were looking for better approaches to reap the benefits of these new developments. This is the time where IT professionals identified major challenges with this approach. Some of them are as follows:

- Scaling the ESB requires scaling all the services implemented in the ESB at once.

- Managing the deployment was difficult since changing one service could impact many other services.

- The ESB approach could not work with the agile development models and container-based platforms.

Most people realized that going forward with the ESB style of networking topology for distributed systems was not capable of gaining the benefits offered by technological advancements in the computing world. This challenge was not only related to ESB, but also to many applications that were developed in a manner where more and more functionalities were built into the same application. The term **monolithic application** was used to describe such applications.

Microservices and containers

This was the time when a set of companies called **Digital Native** companies came from nowhere to rule the world of business and IT. Some popular examples are Google, Facebook, Amazon, Netflix, Twitter, and Uber. These companies became so large that they couldn't support their scale of IT demand with any of the existing models. They started innovating on the infrastructure demand as well as the application delivery demands as their primary motivations. As a result of that, two technologies evolved:

- Container-based deployments
- The microservice architecture

These two innovations go hand-in-hand to solve the problems of increased demand for the aforementioned companies. Those innovations later helped organizations of all sizes due to the many advantages they brought to the table. We will explore these topics in more detail in the upcoming chapters.

Container-based deployments

Any application that runs on a distributed system requires computing power to execute its assigned tasks. Initially, all the applications ran on a physical computer (or server) that had an operating system with the relevant runtime components (for example, JDK) included. This approach worked well until people wanted to run different operating systems on the same computer (or server). That is when virtualization platforms came into the picture and users were able to run several different operating systems on the same computer, without mixing up the programs running on each operating system. This approach was called **virtual machines**, or **VMs**.

It allowed the users to run different types of programs independent from each other on the same computer, similar to programs running on separate computers. Even though this approach provided a clear separation of programs and runtimes, it also consumed additional resources for running the operating system.

As a solution to this overuse of resources by the guest operating system and other complexities with VMs, container technology was introduced. A **container** is a standard unit of a software package that bundles all the required code and dependencies to run a particular application. Instead of running on top of a guest operating system, similar to VMs, containers run on the same host operating system of the computer (or server). This concept was popularized with the introduction of Docker Engine as an open source project in 2013. It leveraged the existing concepts in the Linux operating system, such as **cgroups** and **namespaces**. The major difference between container platforms such as Docker and VMs is the usage of the host operating system instead of the guest operating system. This concept is depicted in the following diagram:

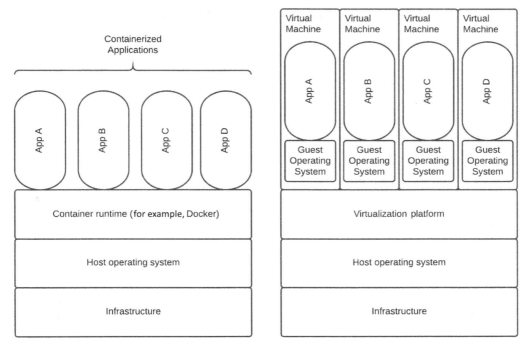

Figure 1.4 – Containers versus virtual machines

The following table provides the key points of distinction between containers and VMs:

Containers	Virtual Machines
Use the same host operating system to run different applications.	Use separate guest operating systems to run different applications.
Portable across different environments and works in the same manner.	Depends on the virtualization technology and operating systems.
Lightweight in size and resource utilization.	Heavyweight in size and resource utilization.

Table 1.1 – Containers versus virtual machines

So far, we've gone through the evolution of the design of distributed systems and their implementation and how that evolution paved the way to the main topic of this chapter, which is the microservice architecture. We'll try to define and understand the microservice architecture in detail in the next section.

What is a microservice architecture?

When engineers decided to move away from large monolithic applications to SOA, they had several goals in mind to achieve the new model. Some of them are as follows:

- Loose coupling
- Independence (deployment, scaling, updating)
- Standard interfaces
- Discovery and reusability

Even though most of these goals were achieved with the technology that was available at the time, most of the SOA-based systems ended up as a collection of large monolithic applications that run on heavy servers or virtual machines. When modern technological advancements such as containers, domain-driven design, automation, and virtualized cloud infrastructure became popular, these SOA-based systems could not reap the benefits that were offered by the same.

For this reason and a few others, such as scalability, manageability, and robustness, engineers explored an improved architecture that could fulfill these modern enterprise requirements. Instead of going for a brand-new solution with a lot of breaking changes, enterprise architects identified the microservice architecture as an evolution of the distributed system design. Even though there is no one particular definition that is universally accepted, the core concept of the microservice architecture can be characterized like so:

> *"The term microservice architecture refers to a distributed computing architecture that is built using a set of small, autonomous services (microservices) that act as a cohesive unit to solve a business problem or problems."*

The preceding definition explores a software architecture that is used to build applications. Let's expand this definition into two main sections.

Microservices are small and do one thing well

Instead of doing many things, microservices focus on doing one thing and one thing well. That does not necessarily mean that it should be written in fewer than 100 lines of code or something like that. The number of code lines depends on many factors, such as the programming language of choice, usage of libraries, and the complexity of the task at hand. But one thing is clear in this definition, and that is that the scope of the microservice is limited to one particular task. This is like patient registration in a healthcare system or account creation in a banking system. Instead of designing the entire system as a large monolith, such as a healthcare application or banking application, we could design these applications in a microservice architecture by dividing these separate functional tasks into independent microservices. We will explore how to break a monolithic application down into a microservice architecture later in this chapter.

Microservices are autonomous and act as cohesive units

This is the feature of the microservice architecture that addresses most of the challenges faced by the service-oriented architecture. Instead of having tightly coupled services, with microservices, you need to have fully autonomous services that can do the following:

- Develop
- Deploy
- Scale
- Manage
- Monitor

Independently from each other, this allows the microservices to adapt to modern technological advancements such as agile development, container-based deployments, and automation, and fulfill business requirements more frequently than ever before.

The second part of this feature is the cohesiveness of the overall platform, where each microservice interacts with other microservices and with external clients with a well-defined standardized interface, such as an **application programming interface (API)**, that hides the internal implementation detail.

Characteristics of the microservice architecture

In this section, we will discuss the different characteristics of a typical microservice architecture. Given the fact that the microservice architecture is still an evolving architecture, don't be surprised if the characteristics you see here are slightly different than what you have seen already. That is how the evolving architectures work. However, the underlying concepts and reasons would be the same in most cases:

- Componentization via services
- Each service has a scope identified based on business functions
- Decentralized governance
- Decentralized data management
- Smart endpoints and dumb pipes
- Infrastructure automation
- Container-based deployments
- Designing for failure
- Agile development approach
- Evolving architecture

Let's discuss these characteristics in detail.

Componentization via services

Breaking down large monolithic applications into separate services was one of the successful features of SOA, and it allowed engineers to build modular software systems with flexibility. The same concept is carried forward by the microservice architecture with much more focus. Instead of stopping at the modularity of the application, it urges for the autonomy of these services by introducing concepts such as domain-driven design, decentralized governance, and data management, all of which we will discuss in the next section.

This allows the application to be more robust. Here, the failure of one component (service) won't necessarily shut down the entire application since these components are deployed and managed independently. At the same time, adding new features to one particular component is much easier since it does not require deploying the entire application and testing every bit of its functionality.

Business domain-driven scope for each service

The modular architecture is not something that was introduced with microservices. Instead, it has been the way engineers build complex and distributed systems. The challenge is with scoping or sizing these components. There are no definitions or restrictions regarding the component's sizes in the architectures that came before microservices. But microservices specifically focus on the scope and the size of each service.

The amount of work that is done by one microservice should be small enough so that it can be built, deployed, and managed independently. This is an area where most people struggle while adopting microservices since they think it is something that they should do right the first time. But the reality is that the more you work on the project, the better you become at defining the scope for a given microservice.

Decentralized governance

Instead of having one team governing and defining the language, tools, and libraries to use, microservices allow individual teams to select the best tool that is suitable for their scope or use case. This is often called the **polyglot** model of programming, where different microservices teams use different programming languages, databases, and libraries for their respective service. It does not stop there, though – it even allows each team to have its own software development life cycles and release models so that they don't have to wait until someone outside the team gives them approval. This does not necessarily mean that these teams do not engage with the experienced architects and tech leads in the organization. They will become a part of the team during the relevant sprints and work with these teams as a team member rather than an external stakeholder.

Decentralized data management

Sometimes, people tend to think that the microservice style is only suitable for stateless applications and they avoid the question of data management. But in the real world, most applications need to store data in persistent storage, and managing this data is a key aspect of application design. In monolithic applications, everything is stored in a single database in most cases, and sharing data across components happens through in-memory function calls or by sharing the same database or tables. This approach is not suitable for the microservice architecture and it poses many challenges, such as the following:

- A failure in one component handling data can cause the entire application to fail.
- Identifying the root cause of the failure would be hard.

The microservice architecture suggests the approach of having databases specific to each microservice so that it can keep the state of the microservice. In a situation where microservices need to share data between them, create a separate microservice for common data access and use that service to access the common database. This approach solves the two issues mentioned previously.

Smart endpoints and dumb pipes

One of the key differences between the monolithic architecture and the microservice architecture is the way each component (or service) communicates with the other. In a monolith, the communication happens through in-memory function calls and developers can implement any sort of interconnections between these components within the program, without worrying about failures and complexity. But in a microservice architecture, this communication happens over the network, and engineers do not have the same freedom as in monolithic design.

Given the nature of the microservice approach, the number of services can grow rapidly from tens to hundreds to thousands in no time. This means that going with a mesh topology for inter-service communication can make the overall architecture super complex. Hence, it suggests using the concept of smart endpoints and dumb pipes, where a centralized message broker is used to communicate across microservices. Each microservice would be smart enough to communicate with any other service related to it by only contacting the central message broker; it does not need to be aware of the existence of other services. This decouples the sender and the receiver and simplifies the architecture significantly. We will discuss this topic in greater detail later in this book.

Infrastructure automation

The autonomy provided by the architecture becomes a reality by automating the infrastructure that hosts the microservices. This allows the teams to rapidly innovate and release products to production with a minimum impact on the application. With the increased popularity of **Infrastructure as a Service (IaaS)** providers, deploying services has become much easier than ever before. Code development, review testing, and deployment can be automated through the **continuous integration/continuous deployment (CI/CD)** pipelines with the tools available today.

Container-based deployments

The adoption of containers as a mechanism to package software as independently deployable units provided the impetus that was needed for microservices. The improved resource utilization provided by the containers against the virtual machines made the concept of decomposing a monolithic application into multiple services a reality. This allowed these services to run in the same infrastructure while providing the advantages offered by the microservices.

The microservice architecture created many small services that required a mechanism to run these services without needing extra computing resources. The approach of virtual machines was not good enough to build efficient microservice-based platforms. Containers provided the required level of process isolation and resource utilization for microservices. The microservice architecture would have not been so successful if there were no containers.

Design for failure

Once the all-in-one monolithic application had been decomposed into separate microservices and deployed into separate runtimes, the major setback was communication over the network and the inevitable nature of the distributed systems, which is components failing. With the levels of autonomy we see in the microservices teams, there is a higher chance of failure.

The microservice architecture does not try to avoid this. Instead, it accepts this inevitable fact and designs the architecture for failure. This allows the application to be more robust and ready for failure rather than crashing when something goes wrong. Each microservice should handle failures within itself and common failure handling concepts such as retry, suspension, and circuit breaking need to be implemented at each microservice level.

Agile development

The microservice architecture demands changes in not only the software architecture but also the organizational culture. The traditional software development models (such as the waterfall method) do not go well with the microservice style of development. This is because the microservice architecture demands small teams and frequent releases of software rather than spending months on software delivery with many different layers and bureaucracy. Instead, the microservice architecture works with a more product-focused approach, where each team consists of people with multiple disciplines that are required for a given phase of the product release.

Evolving architecture

The concepts or characteristics we've discussed so far are by no means set in stone for a successful microservice implementation. These concepts will evolve with time and people will identify new problems, as well as come up with better approaches, to solve some of the problems that the microservice architecture tries to solve. Hence, it is important to understand that the technology landscape is an ever-evolving domain and that the microservice architecture is no exception.

Breaking down a monolith into microservices

Let's try to understand the concepts of the microservice architecture with a practical example by decomposing a monolithic application into a set of microservices. We will be using a healthcare application for this purpose. The same application will be used throughout this book to demonstrate various concepts along the way.

Let's assume we are building an IT system for a hospital to increase the efficiency of the health services provided by the hospital to the community. In a typical hospital, many units exist, and each unit has one or more specific functionalities. Let's start with one particular unit, the **outward patient department** or **OPD**. In an OPD section, a few major functions are executed to provide services to people:

- Patient registration
- Patient inspection
- Temporary treatment
- Releasing the patient from the unit

We'll start with one unit of the hospital and eventually build an entire healthcare system with microservices as we complete this book. Given that there are only four main functions, the IT team at the hospital has developed one web application that covers all these different functional units. The current design is a simple web application with four different web pages, each of which contains a form to update the details captured at each stage with a simple login. Anyone with an account in this system can view the details of all four pages.

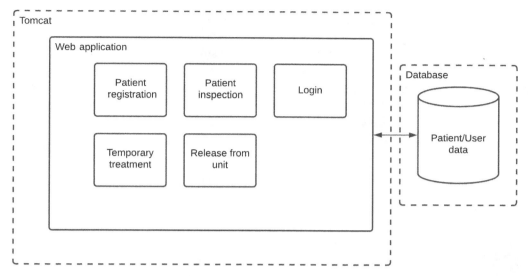

Figure 1.5 – Simple web application for the OPD unit of a hospital

As depicted in the preceding diagram, the OPD web application is hosted in a web server (for example, Tomcat) and it uses a central database server to keep the data. This system works well, and users of this system are given a username and a password to access the system. Only authorized people can access the web application and it is hosted in a physical computing infrastructure inside the hospital.

Let's try to identify the challenges of this approach of building an application as a single unit or a monolith:

- Adding new features and improving existing features is a tedious task that requires a total restart of the application, with possible downtimes.

- The failure of one function can cause the entire application to be useless.

- If one particular function needs more resources, the entire application needs to be scaled (vertically or horizontally).

- Integrating with other systems is difficult since most of the functional logic is baked into the same service.

- It contains large code bases that become complex and hard to manage.

As a result of these challenges, the overall efficiency of the system becomes low and the opportunity to serve more patients with new services becomes harder. Instead, let's try to break this monolithic application down into small microservices.

More often than not, the microservice architecture demands a change in the organizational IT culture, as well as the software architecture and tools. Most organizations follow the waterfall approach to building software products. It follows a sequential method where each step in the sequence depends on the previous step. These steps include design, implementation, testing, deployment, and support. This sort of model won't work well with the microservice architecture, which requires a team that consists of people from these various internal groups and can act as a single unit to release each microservice in an agile manner.

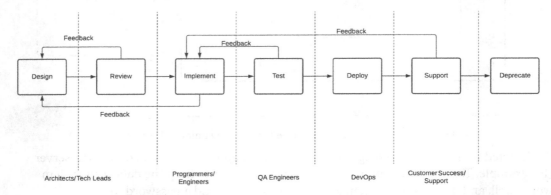

Figure 1.6 – Waterfall development culture

The preceding diagram depicts a typical organizational IT culture where different teams with different expertise (center of excellence or CoE teams) work sequentially to develop and manage the life cycle of a software application. This model poses several challenges, such as the following:

- Longer product release cycles

- Resistance to change causes a lack of frequent innovation

- Friction between teams can cause delayed releases, missing features, and low-quality products

- Higher risk of failure and uncertainty

This kind of organizational culture is not suitable for the highly demanding, innovation-driven enterprise platforms of today. Hence, it is necessary to reduce these boundaries and formulate truly agile teams before starting microservice-style development. Sometimes, it may be difficult to fully remove these boundaries at the beginning. But with time, the individuals and management will realize the advantages of the agile approach.

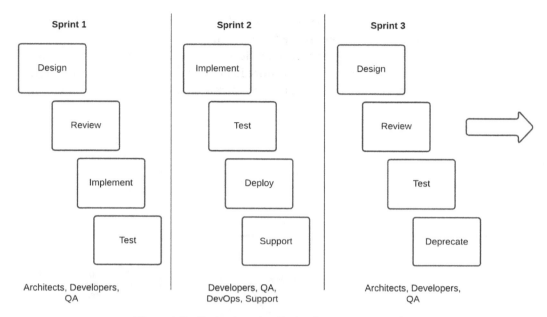

Figure 1.7 – Sprint-based agile development approach

The preceding diagram depicts an approach where the development of microservices is done as **sprints**. These focus on certain aspects of the software development life cycle within a defined time frame. One clear difference in this approach, compared to the waterfall approach, is that the team consists of people from different CoE teams, and a virtual team is formed for the sprint's duration. The responsibility of delivering the expected outcome is on their shoulders.

Let's focus on the application architecture where we identified the challenges with the monolithic approach, which was followed by the OPD web application. Instead of having several functions baked into one single application, the microservice architecture suggests building separate microservices for each function and making them communicate over the network whenever they need to interact with each other.

In the existing design, a common database is used to share data among different functions. In a microservice architecture, it makes sense for each microservice to have its data store and if there is a need to share data across services, services can use messaging or a separate microservice with that common data store instead of directly accessing a shared data store. We can decompose the application into separate microservices, as depicted in the following diagram:

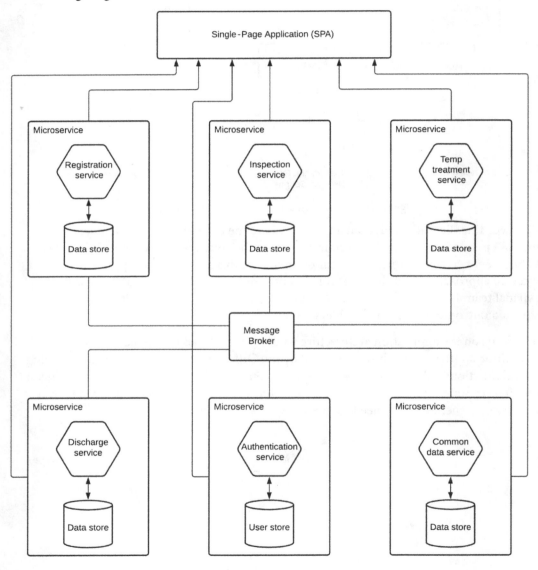

Figure 1.8 – OPD application with microservices

If you find the preceding diagram too complicated, I suggest that you read this book until the end. By doing so, you will realize that this complexity comes with many advantages that outweigh the complexity in most cases. As we can see, the OPD web application is divided into a set of independent microservices that act as a cohesive unit to provide the necessary functionality to the consumers. The following is a list of major changes we made to the architecture of the application:

- Each function is implemented as a separate microservice
- Each microservice has a datastore
- A message broker is introduced for inter-service communication
- The web interface is designed as a single-page application

Let's explore these changes in a bit more detail so that we understand the practicalities of the approach.

Business functions implemented as separate microservices

In the book *Domain-Driven Design* by *Eric Evans*, he explained the approaches to follow when deciding on the boundaries for each microservice with many practical examples. At the same time, he reiterated the fact that this is not straightforward and that it takes some practice to achieve higher levels of efficiency. In the OPD example, this is somewhat straightforward since the existing application has certain levels of functional isolation at the presentation layer (web pages), though this is not reflected in the implementation. As depicted in the preceding diagram, we have identified six different microservices to implement for the OPD application. Each service has a clearly defined scope of functionality, as mentioned here:

- **Registration microservice**: This service is responsible for capturing the details of the patient who is visiting the OPD unit and generates a unique ID for the patient if they have not visited before. If the patient is already in the system, their visit details are updated and they are marked as a patient to inspect.

- **Inspection microservice**: Once the registration is done, the patient is directed to the inspection unit, where a medical officer inspects the patient, updates the details of the inspection results, and provides their recommendation on the next step to take. The next step might involve temporary treatment, discharging the patient from the unit and moving them to the ward unit, or releasing the patient with medication.

- **Temporary treatment microservice**: If the medical officer recommends a temporary treatment, the patient will be admitted to this unit and the required medication will be provided. This service will capture the medication provided against the patient ID and the frequency. After a certain time, the patient will be inspected again by the medical officer, who will decide on the next step.

- **Discharge microservice**: Once the medical officer gives their final verdict on the patient, they will be discharged from the OPD unit to either a long-term treatment unit (ward) or their home. This service will capture the details of the discharge state and any treatment that needs to be provided in case the patient is sent back home.

- **Authentication microservice**: This microservice is implemented to provide authentication and authorization to the application based on users and roles, as well as to protect the valuable patient details from unauthorized access. This service will store the user credentials, along with their roles, and grant access to the relevant microservice based on the user.

- **Common data microservice**: Since the overall OPD unit acts as a single unit that acts on a given common patient, each microservice has to update one common data store for requirements, such as exposing a summary to the external applications. Instead of exposing each service to the external applications, having this sort of common microservice can help with such a function by providing a separate interface that is different from each microservice interface.

Once the microservice boundaries have been defined, the implementation can follow an agile approach where one or more microservices are implemented at the same time, depending on the availability of the resources. Once the interfaces have been defined, teams do not need to wait until another service has been fully implemented. The resources can rotate among teams, depending on their availability. We will discuss additional details regarding the deployment architecture and its implementation details later in this book.

Each microservice has a datastore

One of the main differences between the microservice-based approach and the monolithic approach we discussed in the previous sections is how data is handled. In the monolithic approach, there is a central data store (database) that stores the data related to each section of the OPD unit.

Whenever there is a need for data sharing, the application directly uses the database, and different functions access the same database (same table) without any control. This kind of approach can result in situations where data is corrupted due to an error in the implementation of a certain function and all the functions fail due to that. At the same time, finding the root cause would be hard since multiple components of the application access the same database or table. This kind of design will cause more problems when the applications have a higher load on the database, where all the parts of the application are affected due to the performance of one particular function.

Due to these reasons and many others, the microservice architecture suggests following an approach where each microservice has a data store. At the same time, if there is a need to share a common database across multiple microservices, it recommends having a separate microservice that wraps the database and provides controlled access. If there is a need to share data between microservices, that will be done through an inter-service communication mechanism via messages. We will discuss how to deploy these local data stores, along with microservices, later in this book.

Message broker for inter-service communication

In the monolithic approach, each function runs within the same application runtime (for example, JVM for Java) and whenever there is a need to communicate between functions, it uses an in-memory call such as a method call or function invocation. This is much more reliable and faster since everything happens within the same computer.

The microservice architecture follows a different approach for inter-service communication since each microservice runs on separate runtime environments. Also, these runtime environments may run on different networked computers. Many approaches can be used for inter-service communication, and we will explore all those options in *Chapter 2*, *Why Is Messaging Important in a Microservice Architecture*, of this book. For this initial introduction, we will use the message broker-based approach. We will also be using this throughout this book, so we will discuss it in more detail later in this book.

At the beginning of this chapter, we discussed the different networking topologies and the evolution of distributed systems. There, we identified that having a mesh topology can complicate the overall system architecture and make it harder to maintain the system. Hence, we suggest using a message broker-based approach for inter-service communication throughout this book. The advantages of this approach are as follows:

- A less complicated system
- Loose coupling
- Supports both synchronous and asynchronous communication

- Easier to maintain
- Supports the growth of the architecture with less complexity

We will discuss the advantages of using message brokers for inter-service communication throughout this book.

The web interface is designed as a single-page application (SPA)

As we discussed earlier in this chapter, the microservice style of application development involves making a lot of changes to the way engineers build applications. Traditional web applications are built in a manner where different sections of the application are developed as separate web pages and when the user needs to access a different section, the browser will load an entirely different web page, causing delays and a less optimal user experience.

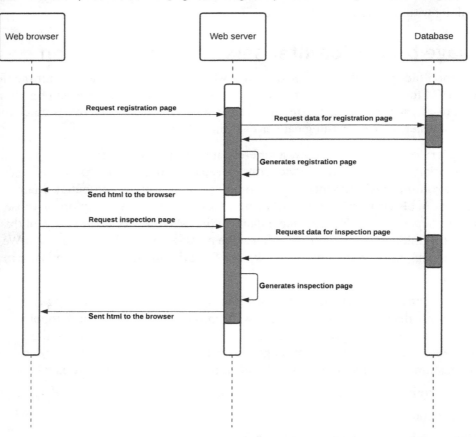

Figure 1.9 – Multi-page OPD web application

As depicted in the preceding diagram, the user is accessing two pages of the web application. Each action triggers the following:

- A request to the web server

- The web server calling the database to retrieve data

- The web server generating the HTML content based on the data

- Responding to the web browser with generated HTML content

These traditional, multi-page applications can cause a significant performance impact to the user due to this multi-step process of loading a web page on the browser.

The concept of SPA addresses these issues and suggests an approach where the entire web application is designed as a single page that will be loaded to the browser at once. A page refresh won't occur when accessing different sections of the application, which will result in better performance and a better user experience.

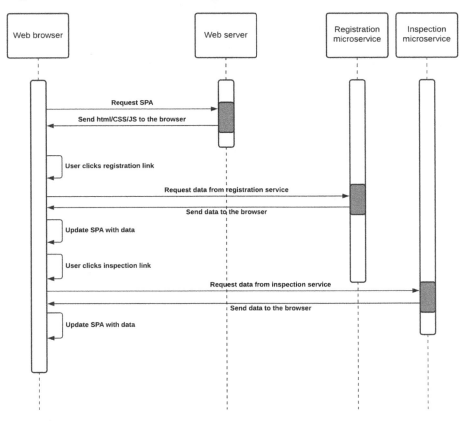

Figure 1.10 – Single-page application with microservices

While there is only a single page for the presentation layer of the application, different microservices implemented at the backend provide the required data to be displayed in the web frontend. The advantage of having separate microservices with an SPA is that users with different privileges will only get access to the authorized details. They will also get access to those details in the shortest possible time since no additional data loading happens.

The preceding diagram only depicts a couple of microservices for the sake of simplicity. The real implementation would interact with all the different microservices we discussed in the previous section.

So far, we've discussed the approach that can be followed when decomposing a monolithic application into a microservice architecture at a very high level while providing a few details on certain aspects. We will be continuing with this topic throughout this book, which means you will get the opportunity to learn more along the way.

Advantages of the microservice architecture

The microservice architecture provides many advantages over its predecessors. Most of these advantages relate to the evolution of the technology landscape. Understanding the evolution of distributed systems helped the microservice architecture become a better approach for building complex distributed systems for the modern era. The following list points out the advantages of the microservice architecture:

- Ability to build robust applications
- Supports the scalability demands of modern businesses
- Better utilization of computing resources
- Helps with innovation with reduced time to market and frequent releases
- Builds systems that are easy to support and maintain

There are several other advantages related to the microservice architecture, but we will start with the aforementioned list and explore more as we continue with this book.

Build robust, resilient applications

The challenge with distributed systems has always been their robustness and their resiliency to failure. The microservice architecture helps organizations tackle this problem by designing applications that are ready for failure and with well-defined scopes. The concept of fail-fast and recovery allows microservice-based applications to identify failures quickly and fix those issues instantly. At the same time, due to the componentized architecture, the failure of one component won't bring down the entire application. Instead, a portion of it and most of the users might not even notice the failure if they don't use that function. This provides a better experience for users.

Build scalable, available applications for modern businesses

The microservice architecture has evolved with the need for scalability in large-scale digital-native organizations, which are required to run thousands of instances of applications and hundreds of different applications. Hence, scalability and availability across wider geographical areas have always been advantages of the microservice architecture. Characteristics such as single responsibility, a modularized architecture, and decentralized data management allow the applications to scale across different data centers and regions without many complications.

Better utilization of computing resources

The popularity of cloud vendors has had a huge impact on the infrastructure costs that are incurred by enterprise software platforms. There were many situations where software systems were utilizing only a fraction of the overall computing infrastructure maintained by these organizations. These reasons paved the way for containers becoming popular and microservices followed the path that was opened up by containers.

Microservices allowed the application to be decomposed into independent units. Each unit can decide on the resources required for it to function. Containers allowed each microservice to define the required levels of resources and collectively, it provided a mechanism to utilize the computing resources in a much better way than the previous monolithic application-driven model.

Helps an innovation-driven organization culture

Modern business organizations are driven by innovations, so having a software architecture that supports that culture helps these organizations excel. The microservice architecture allows teams to innovate and release frequently by choosing the best technology and approach that suits a given business requirement. This urges other teams to also innovate and create an innovation-driven culture within the IT organization.

Build manageable systems

One of the challenges with large monolithic systems was the concept of **Subject-Matter Experts** (**SMEs**) and **center of excellence** (**CoE**) teams, which had control over such applications. Even the **Chief Technical Officer** (**CTO**) would kneel in front of them due to their lack of knowledge regarding those systems. These systems were brittle and the failure of such an application could cause entire organizations to pause their operations. With a defined yet small scope for each microservice and individuals rotating around different teams, microservice-driven applications became much more open to the entire team and no one team had the power to control the system.

Summary

In this chapter, we discussed the concepts of distributed systems and how the evolution of distributed systems paved the way for the microservice architecture, which helps organizations build robust applications with distributed systems concepts. We discussed the key characteristics of the microservice architecture and identified the advantages of it with a practical example of decomposing a monolithic healthcare application into a set of microservices. This chapter has helped you identify the challenges that exist in enterprise software platforms and how to tackle those challenges with microservice architecture principles. The concepts you learned about in this chapter can be used to build scalable, manageable software products for large- and medium-scale enterprises.

In the next chapter, we will get into the nitty-gritty details of building a microservice architecture. We will focus on the important aspects of inter-service communication with messaging technologies.

Further reading

Domain-Driven Design: Tackling complexity in the heart of software, by Eric Evans, available at `https://www.amazon.com/Domain-Driven-Design-Tackling-Complexity-Software/dp/0321125215`.

2

Why Is Messaging Important in the Microservice Architecture?

The microservice architecture operates in a decentralized manner where each function is executed by an independent microservice. At the same time, the success of the overall solution depends on the success of each service, as well as the collaboration among services. This demands a proper messaging framework to handle inter-service communication. Messaging is a method of communication between software components or applications. In a typical enterprise platform, there are two main flows of traffic, namely **North-South** traffic and **East-West** traffic. The North-South traffic represents the communication coming in and going out of a data center from external systems. In the context of microservices, we can replace the data center with a microservice-based platform. The East-West traffic represents the communication within the data center. For microservices, we can define East-West traffic as the communication within the microservice platform between individual microservices. These two different message flows have different requirements and expectations that need to be met by the messaging model and the platform that has been selected. We will explore the suitable messaging architectures for both types of traffic patterns within this chapter.

In this chapter, we will explore the technologies that are used when designing and implementing a distributed system and focus on the topic of messaging-based communication in a microservice architecture.

We are going to cover the following main topics in this chapter:

- Messaging patterns used in distributed systems
- Understanding the communication requirements of a microservice architecture
- Common messaging technologies used in a microservice architecture

Messaging patterns used in distributed systems

As we saw in the previous chapter, distributed system designers use different networking topologies to interconnect systems, and most of those topologies are still used heavily in the technology industry. Network topology defines the layout of communication within the platform and the actual communication pattern; the protocols are not well defined there. In this chapter, we will discuss the communication that occurs within a distributed system.

Distributed system designers have produced different approaches to communicating between systems. These patterns include the following:

- **Remote procedure calls** (**RPC**) versus shared resources
- Synchronous versus asynchronous (client-server versus pub-sub)
- Orchestration versus choreography

Let's look at the preceding approaches in more detail so that we can use them when designing communication patterns for a microservice architecture, which is another distributed system at its core.

RPC versus shared resources

When two systems are needed to share data, using a shared resource such as a database or a file server is one of the simplest options available. This model of communication requires a third system to bridge the communication gap between the source system and the target system and acts as a hub:

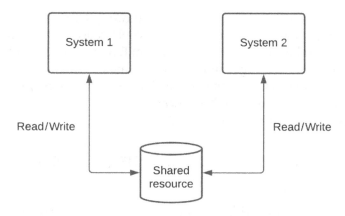

Figure 2.1 – Shared resource-based communication

Each communication goes through a shared resource, and the source and target systems do not necessarily need to be aware of each other's existence if there are no issues with data processing. This leads to the fundamental problem of concurrent access, where both systems are trying to access and/or modify the same data (table/row or file/line) at the same time. System designers and engineers have developed several solutions to this concurrency problem via mechanisms such as coordination, locks, semaphores, and so on. Yet, it is one of the most challenging problems in distributed system design.

Instead of using a third system to communicate with each other, designers produced the idea of remote procedure calls, where the source system directly communicates with the target system by executing a remote procedure call or remote function call over the wire. The source system executes the function while assuming it is running on the source system. But the actual function executes a remote system, which is the target system:

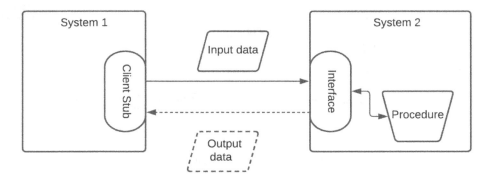

Figure 2.2 – Communication using a remote procedure call

This approach requires a good understanding between two systems so that they have tight coupling. Making a change to the target system could cause a lot of problems for the other calling systems and hence to the overall solution. Engineers and architects worked through these fundamental challenges and produced standard interfaces so that the source system and the target system could execute independently if they adhered to these standards. Some examples are **Simple Object Access Protocol (SOAP)**, **Common Object Request Broker Architecture (CORBA)**, and **gRPC**.

Synchronous versus asynchronous (client-server versus event-based)

When two systems communicate with each other, the simplest mode of operation is the synchronous communication model, where the system that initiates the communication (source system) waits for a response from the other system (target system) before sending another message.

The following diagram depicts the synchronous communication model, where the source system sends a message in the form of a request and waits until the target system comes back with a response:

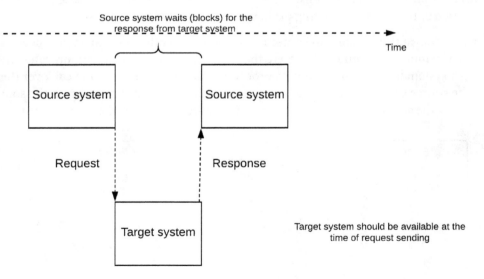

Figure 2.3 – Synchronous messaging model

This model is the most used messaging model in the world wide web or the internet, where users request a website and get the website (HTML representation) as a result on their computer. This is displayed by a web browser in a nicely formatted manner. This synchronous model is often called the **client-server model**, where the initiator of the message becomes the client while the responder becomes the server. This model is not complicated to implement on both the client side as well as the server side. Hence, it has become the most popular messaging model so far.

But not all communication between systems works well with this model. Especially when there are systems that require high performance, parallel processing, and advanced guarantees on data, this model does not work. The reason for this is that the client must block (wait) until it gets a response from the server, and this will impact the overall performance and throughput of the system. Also, both the client and the server must be available during the time of sending messages to make successful communication. Hence, the asynchronous communication model is used in such scenarios.

The following diagram depicts the asynchronous communication model, where the source system and the target system communicate over an intermediate message broker:

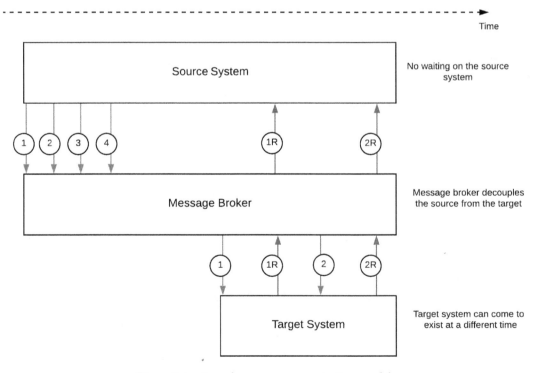

Figure 2.4 – Asynchronous communication model

In asynchronous communication, the sender does not wait for a response from the receiver and, in most cases, does not care about or is not aware of the receiver's existence. The sender will send (publish) the message to an intermediate broker (message broker) and once this broker responds with an acknowledgment, the sender will move on to the next task. Because of this approach, there is less waiting on the sender's side and the overall performance of message sending is improved.

At the same time, the recipients of these messages don't need to always be available when the sender sends the message. Depending on the broker technology and the publishing model being used by the system, messages can be stored for a longer time until the recipient is available in case the recipient is not available at the time of message delivery. This model is also called the **publish and subscribe (pub-sub)** model, where the sender becomes the publisher who publishes a message to an entity called a **topic** or a **queue**, and the recipient of the message becomes the subscriber who subscribes to the same topic or queue.

There are two main variants of asynchronous communication models and depending on the message broker that you choose, those brokers will offer these models with their own implementations. But the core concepts are the same for the users:

- Queue-based
- Topic-based

The queue-based model is suitable for scenarios where a message needs to be sent to one and only one subscriber out of many interested subscribers. The workload can be distributed to multiple subscribers so that they can process them independently. This is like load balancing the messages across multiple subscribers.

The following diagram depicts the queue-based asynchronous communication model, where a message is delivered to one of the interested subscribers:

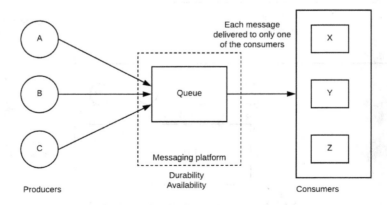

Figure 2.5 – Queue-based asynchronous communication

The topic-based model is suitable for scenarios where each message needs to be distributed to multiple recipients who have an interest in the message. A perfect example of this model is distributing news from sources across different news channels:

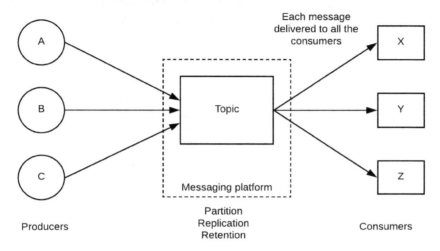

Figure 2.6 – Topic-based asynchronous communication

In this section, we discussed the fundamentals of the asynchronous communication model. More details about the usage of this model will be covered in later chapters.

Orchestration versus choreography

In the previous two sections, we discussed how two or more systems within a distributed system communicate with each other using a shared entity. Sometimes, this entity is a shared resource such as a database or a file server, while in other scenarios, it is a message that is transferred from one system to the other. We have not discussed anything other than a single communication request so far, so this is what we will be discussing in this section.

One of the most important aspects of distributed system design and implementation is how the overall system works as a cohesive unit to produce great experiences for the consumers of the platform. There will be many use cases where an end-to-end user story involves a series of interactions to produce the expected outcome.

The **orchestration approach** involves a central system or a component that coordinates the communication between multiple systems to produce the expected result. A practical example of orchestration is a musician who directs an orchestra in a musical. It follows a sequential execution approach where the coordinating system will operate synchronously with target systems, execute a sequence of such interactions, and produce a final response:

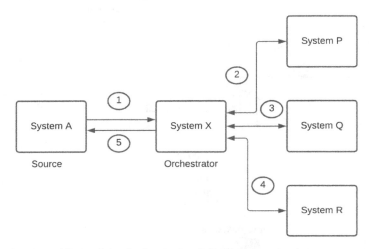

Figure 2.7 – Orchestration-based communication

As depicted in the preceding diagram, three steps are being followed:

1. The source system initiates communication with the orchestrator system.
2. The orchestrator delegates the task to three other systems in a sequential manner and produces the result.
3. The orchestrator shares that back to the source system.

The actual execution of delegated tasks can occur either in a sequential or parallel manner, depending on the use case. The most important characteristic of this model is that there is a central coordinator that controls communication flow. The orchestration-based approach is used in enterprise use cases such as enterprise service bus and **service-oriented architecture** (**SOA**)-based application integration.

Choreography is a technology that's used in dancing where a team of dancers acts as a single unit to produce a *show* to the audience. There is no central coordinator or a person who controls the dance, but there is a choreographer who designs the dance before the show, makes sure dancers do their act properly during the show, and collaborate at the time of the show. Distributed systems designers have identified this approach as a possible solution for building highly scalable, highly distributed systems that can scale to thousands of distributed components:

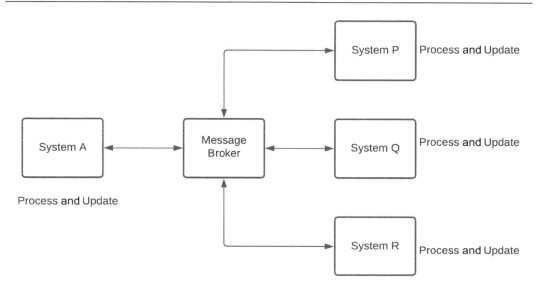

Figure 2.8 – Choreography-based communication

One of the fundamental differences between the choreography that's used in dancing and distributed systems is that the latter requires a messaging component to act as the bridge between actors, while the former does not require an additional layer for such communications since that happens through their eyes and ears. As per the preceding diagram, all the systems operate independently, and those systems will act on certain events that are received via the message broker, process these events, and update the relevant topics or queues with the response. The central message broker acts as the communication medium but not as the coordinator, which we saw in the *orchestration model*.

Communication requirements of a microservice architecture

The microservice architecture is a distributed system. The communication from and to these individual services plays a pivotal role in the overall design and implementation of the platform. Before discussing any particular messaging model for the microservice architecture, let's try to understand the communication requirements of a microservice architecture.

Let's start with the example use case we discussed in the previous chapter, where we started designing a solution for a hospital **outward patient division** (**OPD**) using the microservice architecture. Four major functional units operate within the OPD, and each of these functional units can operate independently while communicating with each other. In addition to the functional unit-based definitions of microservices, we also derived two additional microservices for authentication and data sharing, as depicted in the following diagram:

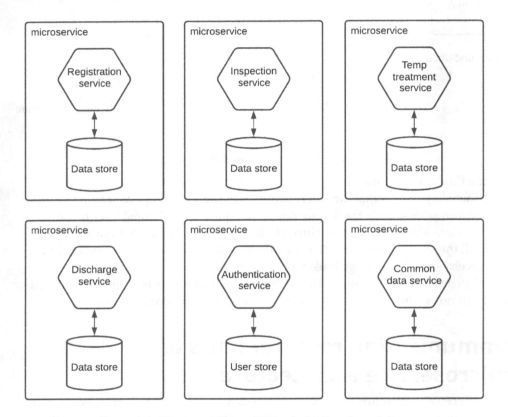

Figure 2.9 – Decomposing an OPD solution as microservices

Once the solution has been decomposed into domain-driven microservices, as we did here, the next step is to define how they interact with each other (East-West traffic) and with external consumers (North-South traffic). In this example, North-South (also known as North-bound) traffic is received from the web application's frontend (for example, a single-page application) that is used by the hospital staff. This interaction needs to occur in real time in most cases, and using synchronous-style messaging would suffice here.

As an example, the patient will be registered by one of the hospital staff members and they will access the web application and update the details about the patient in the relevant UI section:

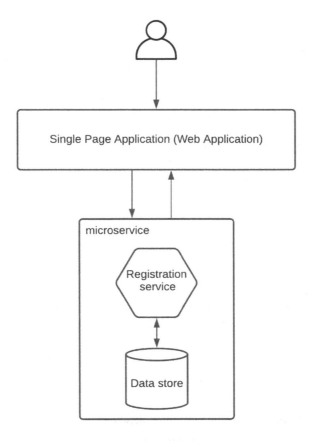

Figure 2.10 – Interaction between the web application and the registration service

This interaction needs to be done in real time, and the user (staff member) should get a response from the service with a possible patient ID that can be used for the rest of the functions for a given patient. This requirement can be supported with a communication model such as **Representational State Transfer** (**REST**). At the same time, this generated patient ID and their details need to be stored via the common data service so that other functions (services) can access these details to treat this patient. If we design this inter-service communication using the same synchronous model, then the common data service needs to be running at the time of the registration, and any failure on that service will impact the patient registration functionality. This is not a clever design since the failure of an internal service (common data service) should not cause failure on the entire system.

Hence, we can use an asynchronous communication model when communicating from the registration service to the common data service. This ensures that the unavailability of one service will not cause any issues on the other service, as well as on any other parts of the system:

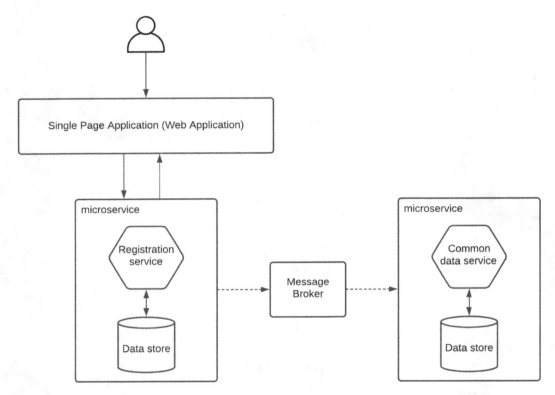

Figure 2.11 – Asynchronous communication between the registration service and the common data service

With this mode of communication, the registration service can operate independently without worrying about the status of the common data service. The intermediate message broker takes care of the delivery guarantee based on the needs of the users of the system.

Once the patient has been registered, that patient will be redirected to the inspection unit, where they need to be inspected before being given any medication. When the patient moves from the registration section to the inspection section, the doctor would like to receive the patient information instantly rather than going to the system to search for the patient's ID. This requirement can be achieved with the event-based communication model with a message broker.

Once the patient has been registered, the patient's details are sent to the message broker so that the common data service is updated with that information. The same event can be delivered to the inspection service via a common topic. At the inspection section, it can use its own business logic to process the event and use the required details:

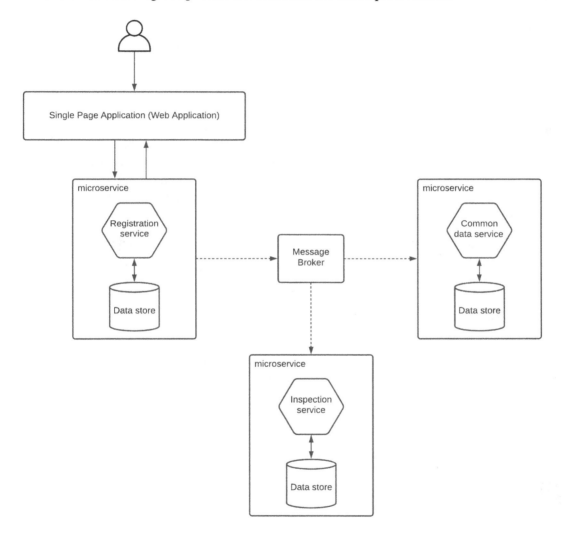

Figure 2.12 – Microservice communication via a message broker

The preceding diagram depicts how the registration service shares the patient details to both the common data service and inspection service via the message broker asynchronously. This approach can be utilized when communicating with the rest of the services as well.

Common messaging technologies used in a microservice architecture

We discussed how message brokers can be used in a microservice architecture for East-West traffic in the preceding section. At the same time, we discussed how a synchronous communication model can be utilized for North-South traffic. We will explore some of the popular approaches that are used in the industry for both North-South and East-West traffic in detail in this section.

North-South traffic

In a typical enterprise platform, North-South traffic is the more sensitive and customer-facing communication flow that attracts the maximum attention from everyone. At the same time, it has a direct impact on the business. Since this message flow happens between a human user and a system, a synchronous communication model is used in most cases. Certain requirements need to be met when selecting a technology for North-South communication. This includes the following:

- **Security**: Make sure only authorized people have access to business data.
- **Performance**: Users should have a better experience when consuming the services.
- **Monitoring**: Need to identify the usage behaviors that drive business innovations.
- **Control**: The incoming traffic should be controlled so that the best experience is delivered to users.

The following is a list of common techniques that are used for North-South traffic that can fulfill the requirements mentioned in the preceding section:

- REST
- GraphQL
- WebSocket

REST provides a simple yet powerful communication model for messaging synchronously. The client interacts with the server using a **Universal Resource Identifier (URI)** and a method that defines the operation that is going to be executed at the server. This interaction can be predefined with a definition such as **Swagger** or **OpenAPI Specification**, which hides the implementation details. Many tools have been built for implementing REST-based systems, including API gateways. These tools provide the required security, performance, scalability, and availability for the enterprise when using REST-based approaches.

With the increased popularity of REST in the technology world, some of the limitations it had were overlooked by common users. One such limitation is that when using REST, users cannot specify the amount of data that is received from the client. This is sometimes called **over-fetching**. If the backend contains thousands of sets of data in the response, a client has no choice but to accept the entire payload. GraphQL addresses this concern by providing control to the client with a query language type syntax, where a client can specify how the data is returned. A GraphQL client can choose the fields of data to be returned, or even rename the response objects as they deem to be returned to them. While REST focused on simplicity and ease of usage, GraphQL focused on performance and stability. Several tools are being built to implement GraphQL-based solutions in the industry. GraphQL also has a schema that defines the interface for client interactions.

WebSocket has evolved as an alternative approach to HTTP 1.1 (which was the underlying protocol for REST), which had the limitation of building proactive web applications due to the half-duplex type communication model it was built on. WebSocket allowed servers to communicate with the client without an actual request for data. This allowed users to build web applications such as real-time monitoring dashboards, where those dashboards were automatically updated with data coming from the server itself (this is called a server push). It also improved the overall performance of the communication by reducing the overhead of sending headers per message and the usage of frames.

These technologies provide options for microservice developers to choose from based on their needs. There are tools such as API gateways that support all these technologies and expose services to external consumers:

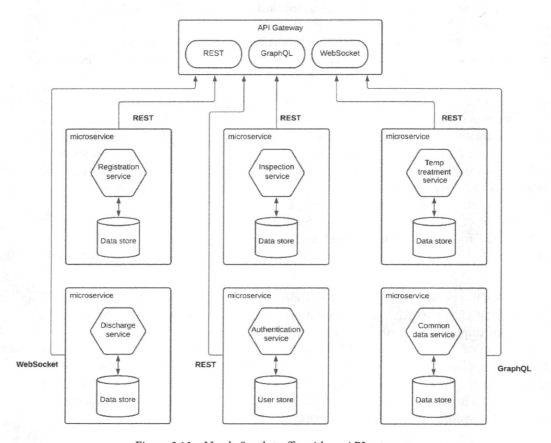

Figure 2.13 – North-South traffic with an API gateway

The preceding diagram depicts how a given microservice architecture can implement different messaging protocols for North-South traffic. It also shows that a common API gateway can be used to aggregate these services and expose them to the consumers. This API gateway could apply certain functionality such as security, rate limiting, and monitoring for the underlying microservices.

East-West traffic

Most of the time, architects and technical leads give priority to the consumer-facing interfaces and they let developers decide on the internal communication between services, which is the East-West traffic. In a monolithic application, East-West traffic goes within the program itself inside the memory. But in a microservice type architecture, East-West traffic also goes over the network. That is why it is important to consider East-West traffic as a high priority.

Certain aspects need to be addressed by the technology when it comes to inter-service communication or East-West traffic. Some common capabilities include the following:

- **Resiliency**: Ability to withstand failures.
- **Performance**: Should be able to handle a larger communication network.
- **Robustness**: Clear interfaces to communication.
- **Loose coupling**: Services should not be coupled with each other.

When the monolithic applications were decomposed into a modularized architecture with **Service-Oriented Architecture (SOA)**, developers used the same communication method for both North-South and East-West traffic, which is HTTP (REST and SOAP). Then, microservices came in and decomposed these applications further with hundreds/thousands of *micro components* that need to communicate with each other. That is when the developers and architects realized that HTTP would no longer be the best possible solution in such scenarios. The following are some technologies that are used for East-West traffic in a microservice architecture:

- HTTP (REST)
- NATS
- Kafka
- gRPC

The simple nature of REST over HTTP and the experience that developers had with the same made it the first choice for East-West traffic in a microservice architecture. Due to its popularity and adoption, there are many tools and libraries available to achieve the requirements of East-West traffic, as discussed in the preceding section. The synchronous nature of the REST over HTTP model made it unsuitable for use cases where an asynchronous communication model is required.

That is when the message broker-based asynchronous communication model started to become useful in East-West communication. It allowed services to communicate as and when required without worrying much about the existence or the availability of the consumer. This model of asynchronous communication provides certain benefits, such as the following:

- **Improved performance**: Able to handle a higher number of messages at any given moment since there is no dependency on the consumer.

- **Effective message delivery**: Consumers can receive messages when they want and higher levels of guarantees are provided.

There are many different message broker tools available in the market that can support the requirements of East-West communication. We have picked two popular tools (NATS and Kafka) to look at here.

NATS

NATS is a simple and secure messaging platform that allows developers and architects to build effective distributed applications. It provides the following features:

- High performance
- Lightweight
- At most once and at least once delivery
- Always on and available
- Supports both the request-response and pub-sub messaging models

NATS supports microservice communication (East-West traffic) with a message broker-based approach. We will discuss NATS in detail in *Chapter 3, What Is NATS Messaging?*.

Kafka

Kafka is another popular messaging framework for building message broker-based distributed systems. It has a unique yet simple architecture that can integrate well with existing enterprise applications. It decouples the messaging producers from the message consumers while providing the reliability of message delivery at scale:

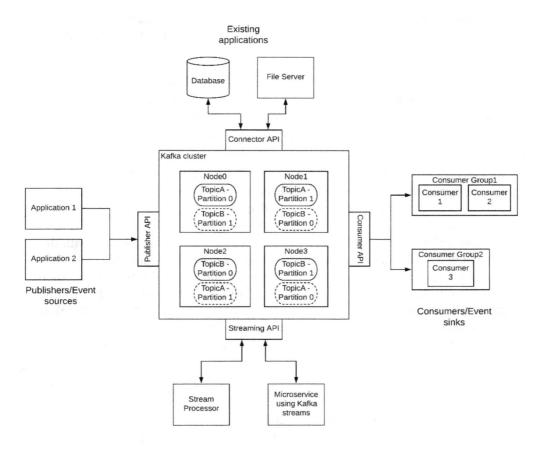

Figure 2.14 – Kafka architecture

The preceding diagram depicts the Kafka architecture, along with the APIs provided by Kafka for integration. Kafka can run on a cluster of nodes and topics that are divided into partitions and replicated across different nodes to provide higher availability and better message guarantees. It has a leader election mechanism where a certain node becomes the leader that holds a given partition and provides access to the publishers and consumers. The other nodes that hold the same partition become the followers who have a replica of the data.

Kafka can expand across geographically distributed resources. Once a Kafka cluster has been set up, external applications and systems can interact with the cluster through four standard APIs:

- **Publisher API**: This is used by message producers to publish messages to the cluster. These messages are stored for a configurable duration, regardless of whether they are consumed. When the events are being published, publishers can specify the partition in which the event is going to be stored in each topic.

- **Consumer API**: The events that are stored in the topics are delivered to consumers who are subscribed to these topics. Unlike in standard pub-sub-based topics, the consumer has total control over reading the messages from the topics. The consumer can control the offset value, which tells us about the position of the message within a given partition so that consumers can read messages from any location within a given topic and a partition. Additionally, Kafka comes with the concept of a consumer group, through which the consumers can balance loads across multiple competing consumers, such as a queue-based subscription. There can be multiple consumer groups subscribed to a given topic, and each consumer group will get one copy of the message at a given time. A given consumer within a consumer group is bounded to a given partition within a topic. Hence, there cannot be more consumers within a group than the number of partitions of a topic.

- **Connector API**: Sometimes, Kafka needs to be integrated with existing applications and systems. As an example, if we want to read a set of log files through Kafka to process the data within those files using a stream processor, Kafka provides the connector API to implement reusable components called connectors. The same connector can be used for any file-based integration with Kafka.

- **Streaming API**: The events that are stored in Kafka sometimes need to be processed in real time for various purposes such as transformation, correlation, aggregation, and so on. These sorts of functionalities can be applied through a stream processor, which acts on a continuous stream of records. The Kafka streaming API exposes a topic as a stream of events so that any other stream processing engine can consume that and do the processing.

Kafka is used by many enterprises to build event-driven applications. The following are some use cases that Kafka can be used for:

- Reliable messaging platforms that deliver messages across systems at cloud scale.

- Real-time event processing systems with machine learning and AI-based processing.

- Large-scale data processing systems such as log monitoring and ETL-based systems.

- Build systems with event sourcing where each activity is considered an event that is attached to an event log.

gRPC

gRPC is a modern **remote procedure call (RPC)** framework developed by Google for messaging. It is designed to connect services in and across data centers with support for load balancing, tracing, health checking, and authentication. In addition to these core capabilities, some of its advantages include the following:

- **Simple service definitions**: You can define the service messages using protocol buffers, a powerful binary serialization toolset.

- **Bi-directional communication**: Supports full-duplex communication over the HTTP/2 protocol.

- **Portable**: Supports generating clients and servers from the service definitions for multiple languages.

- **Scalable**: Start small and scale to global-scale traffic loads.

In a typical RPC system, the client calls a method in the server, assuming it is running on the client side within the same server. The service definition contains details about the methods and parameters that need to be passed in when making the call:

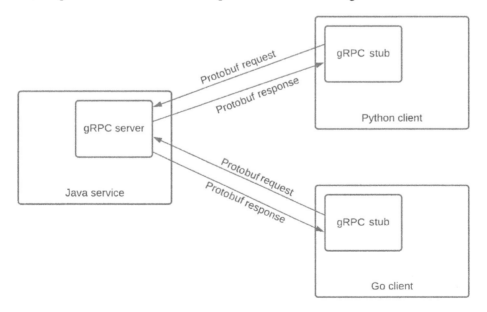

Figure 2.15 – gRPC architecture

As depicted in the preceding diagram, gRPC clients and servers can run and communicate with each other in a variety of environments and can be written in any of the supported languages. Even though RPC-style communication is used primarily in synchronous communication models, gRPC programming APIs support both synchronous and asynchronous methods.

Summary

In this chapter, we discussed the importance of messaging within a distributed system and how it relates to the success of the platform. We started by identifying different messaging models used in distributed systems and discussed these models in detail. Then, we dived into the OPD example that we discussed in the previous chapter and identified some common messaging requirements of a microservice architecture.

We covered two types of communication flows within a microservice architecture and discussed how each type can be implemented with the technologies available on the market. Understanding messaging-based communication helps developers and architects choose the best technology and method when building real-world enterprise platforms.

We'll discuss the NATS messaging framework in detail in the next chapter and cover various concepts, examples, and use cases surrounding this topic.

3
What Is NATS Messaging?

Modern enterprise applications are distributed in nature. These distributed applications require a communication medium to share data between components. Messaging is the approach that distributed systems use for this purpose. NATS is a modern messaging framework that allows developers and operators to build distributed systems. NATS makes it easier for programs to communicate across different environments, languages, cloud platforms, and on-premises systems. NATS is designed to meet the needs of modern distributed systems while supplying a secure and simple API for users. It is used in real-world applications such as the following:

- Microservice-based applications:

  - Service mesh and inter-service messaging

- Event streaming applications:

  - Observability/analytics

  - Machine learning

  - Artificial intelligence

- Command and control-based systems:

Internet of Things (IoT) and edge computing

- Telemetry/sensor data/command and control

- Augmenting or replacing legacy messaging systems

In this chapter, we will explore the NATS messaging technology in detail by going through its features, concepts, and use cases. We will be covering the following main topics:

- History of NATS

- Introduction to NATS

- How does the NATS protocol work?

- Advanced NATS concepts

- Advantages of NATS

By the end of this chapter, you will have gained a detailed understanding of the NATS messaging framework and how it is used in real-world applications.

Technical requirements

In this chapter, we will install the NATS server and configure it to demonstrate certain aspects of the NATS messaging system. Hence, you should have a basic understanding of using the command-line tools based on the operating system you are using. As a prerequisite, you can install the NATS server by following the instructions mentioned in the official NATS documentation: `https://docs.nats.io/nats-server/installation`. We will be using NATS server version 2.x (v2.2.2) to demonstrate the examples in this chapter.

History of NATS

NATS messaging was initially developed for the Cloud Foundry platform as the message bus that was used by internal components to share data. It was initially written in the Ruby programming language since most of the Cloud Foundry components were developed using the same language. The Ruby-based implementation was performing well enough with over 75,000 messages per second being processed by a single NATS server running on commodity hardware. But NATS creator *Derek Collison* wanted to go even faster.

That is when he decided to rewrite the NATS server and the client using the Go programming language in 2012. This version of the server was processing 2 million messages per second, and it kept improving with time. By now, it can process around 18 million messages per second. NATS also supports cloud-native systems such as microservices and works well with containers and container orchestration platforms such as Kubernetes. Due to the cloud-native support provided by the NATS framework, it was accepted by the **Cloud Native Computing Foundation** (**CNCF**) as a project to be hosted there.

Introduction to NATS

NATS is a messaging platform built to meet the distributed computing needs of modern applications. It addresses distributed computing needs such as the following:

- Secure communications between services
- Service discovery
- Resiliency
- Agile development
- Scalability
- Performance

These requirements are fulfilled by the unique design of the NATS messaging framework, which comes with features such as the following:

- **High performance**: NATS performs better than most of the existing message broker products, including Kafka and RabbitMQ.
- **Lightweight**: It does not need sophisticated hardware and complex deployment models to support large message volumes.
- **Simple to use**: It supplies a simple API to use the system.
- **At most once and at least once delivery**: Supports message guarantees required by applications.
- **Support for event handling**: It can handle event streams with better performance.
- **Support for different languages**: NATS has clients for over 30 programming languages.

At its core, NATS enables applications to exchange data in the form of messages. These messages are addressed by subjects and do not depend on the underlying network. Let's try to understand the different messaging models supported by NATS. All the different messaging models are built on top of the fundamental publish and subscribe messaging model, which is asynchronous. NATS supports the following four main messaging models:

- Subject-based messaging
- Publish-subscribe
- Request-reply
- Queue groups

Let's discuss each of these messaging models in detail.

Subject-based messaging

This messaging model works on top of the topic-based publish-subscribe approach we discussed in *Chapter 2, Why is Messaging Important in the Microservice Architecture?*. In this model, the messages are scoped with a **subject** so that any interested party can subscribe to the messages based on it. In simple terms, a subject is a string that forms a name that both the publisher and subscriber use to share messages.

The following diagram shows the subject-based messaging model, where a publisher publishes a message with the `patient.profile` subject and the subscribers are subscribed to the same subject:

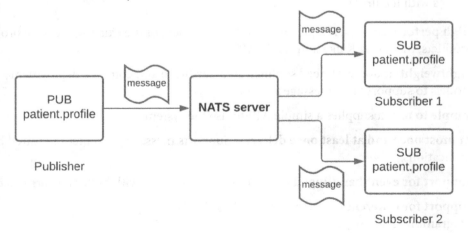

Figure 3.1 – Subject-based messaging with NATS

In the preceding diagram, both the publisher and the subscribers are using the same name as the subject. The NATS server supports alphanumeric characters and the . (dot) character for subject names. These subject names are case-sensitive and cannot contain whitespaces. The messaging approach depicted in the preceding diagram is like the topic-based publish-subscribe model. The real advantage of subject-based messaging can be obtained by using the subject hierarchy concept of using the . character. The following subject names create a hierarchy for patient information:

- `patient.profile`
- `patient.profile.name`
- `patient.profile.address.city`
- `patient.profile.bloodtype`

By using the aforementioned subject hierarchy and the wildcard characters supported by NATS, subscribers can subscribe to the messages selectively without specifying the exact subject name. NATS supplies two wildcard characters that can be used in the subject when subscribing to the messages. Subscribers can use these wildcard characters to listen to multiple subjects with a single subscription. But the publishers must always use the full subject name without the wildcard characters.

The first wildcard is `*`, which matches a single token. As an example, if an application needs to listen for patient names, it could subscribe to the `patient.*.name` subject, which would match `patient.profile.name` and `patient.first.name`. This concept is depicted in the following diagram:

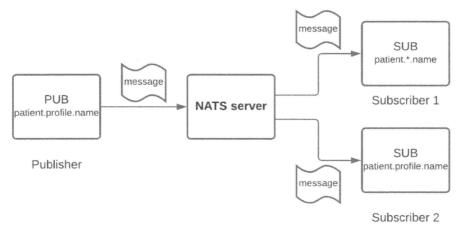

Figure 3.2 – Subject-based messaging with a single token wildcard in NATS

The publisher is publishing the message with the full subject name, while **Subscriber 1** has subscribed to the subject using the wildcard (`patient.*.name`) and **Subscriber 2** has subscribed to the subject using the full subject name. Both subscribers will get the message in this scenario.

The other wildcard choice is >, which will match one or more tokens in the subject, and it can only appear at the end. As an example, if an application needs to listen to all the messages related to the patient, it could subscribe to `patient.>`, which will receive the messages that have been published by the `patient.profile`, `patient.profile.name`, and `patient.profile.address.city` subject names. The following diagram depicts an example of a wildcard subscription:

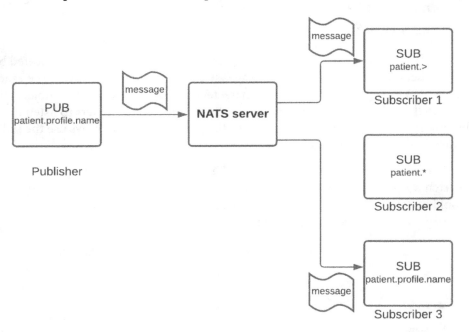

Figure 3.3 – Subject-based messaging with a multi-token wildcard in NATS

The preceding diagram depicts how the multi-token wildcard, >, matches the message that was sent with the `patient.profile.name` subject while the single token, *, misses the same message. If an application needs to listen to all the messages that have been published to all the topics, then it can subscribe to > as the subject name, as depicted in the following diagram:

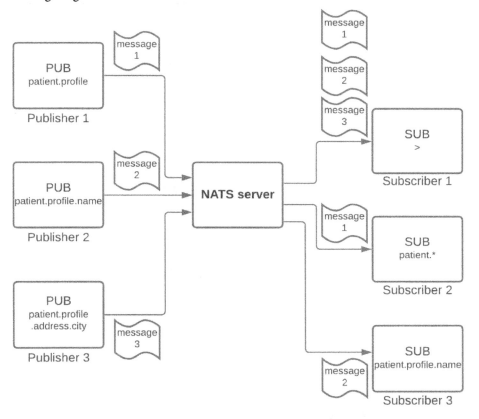

Figure 3.4 – Subject-based messaging with the wiretap wildcard option in NATS

The preceding diagram depicts a use case where three applications are publishing messages with three different subject names and three subscribers are listening to different subject names. The first subscriber listens to the > subject name, which will receive all the messages that have been published with all the subject names. The second subscriber listens to the `patient.*` subject name and will receive the message that's been published with `patient.profile`. At the same time, the third subscriber listens to the `patient.profile.name` subject name, which receives the message that was published with the same name. This approach of listening to all the messages can be used for monitoring and tracing purposes. It is sometimes called the **wiretap** pattern in messaging.

Publish-subscribe messaging

The most common messaging model supported by message brokers such as NATS is the **publish-subscribe** or **pub-sub** model, where a message is published by an application (publisher) to a topic and a set of other applications (subscribers) that are interested in the message subscribe to the topic and receive the same message. This model can be seen in the following diagram:

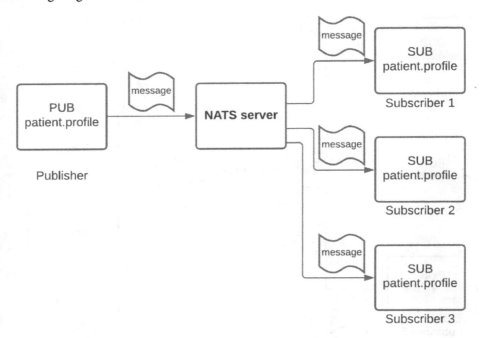

Figure 3.5 – Publish-subscribe messaging with NATS

The preceding diagram depicts the pub-sub messaging model in NATS, with the usage of the subject in place of a topic. The publisher sends the message with a given subject name and all the subscribers are subscribed to the same subject. This model is a subset of the subject-based messaging options we discussed earlier. This pattern is called the **fan-out** or **one-to-many messaging pattern** as well.

Request-reply messaging

This is the communication or messaging pattern that's used in most distributed applications and systems. In this model, a source system (requester) sends a message to a target system (responder) and expects a message in response. In most cases, this method is implemented in a synchronous manner, where the requester will wait for the response before doing anything else. In the world of NATS, it is implemented in an asynchronous manner using the publish-subscribe model with the usage of two subjects.

A publisher sends a message on a particular subject by attaching a reply subject name along with it. Then, the subscribers who receive this message send the response to the reply subject that was specified in the request message. Reply subjects have a unique name called **inbox** and the NATS server will send the messages coming into these subjects to the relevant publisher. This model can be seen in the following diagram:

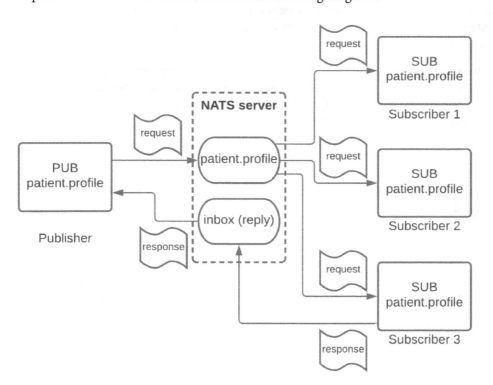

Figure 3.6 – Request-reply messaging with NATS

The request message is published to the `patient.profile` subject and the subscribers are listening on the same subject. The third subscriber responds with a response message that will be routed back to the publisher by the NATS server. NATS allows multiple subscribers to form a queue group so that only one subscriber responds to each message and distributes the load at the receiving end. In addition to that, NATS also allows multiple responses, where the first response is used and the system efficiently discards the added responses. This allows applications to run with the best response latency from a set of responders that are running independently.

Queue groups messaging

In a scenario where message processing needs to be distributed across several target systems, NATS provides the built-in load balancing capability with queue groups. Using this model, subscribers will receive messages in a load-balanced manner, and it will help the system distribute the workload. Each subscriber who needs to listen to a queue group needs to register for the same queue name.

When messages are published with the registered subject name, one member of the group is randomly chosen and receives the message. Although queue groups have multiple subscribers, each message is received by only one subscriber. The following diagram shows how a queue group can be used with NATS:

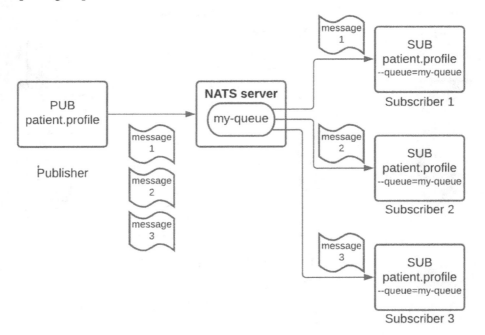

Figure 3.7 – Queue groups messaging with NATS

The preceding diagram shows a use case where a publisher application is sending three messages named **message 1**, **message 2**, and **message 3** to the `patient.profile` subject. These messages are received by three separate subscribers listening on the same subject name with the `my-queue` queue name. One important thing to note here is that the publisher is not aware of the queue. Only the subscribers will be aware of the queue name and refer to that when listening to the subject name. Queue group names follow the same naming convention as subjects. Foremost, they are case-sensitive and cannot contain whitespace.

One of the excellent features of NATS is that queue groups are defined by the application and their queue subscribers, not on the server configuration.

Queue subscribers are ideal for *scaling services*. Scaling up is as simple as running another application, while scaling down includes stopping the application with a signal that drains the inflight requests. This flexibility and lack of any configuration changes makes NATS an excellent service communication technology that can work with all platform technologies.

How does the NATS protocol work?

At the beginning of this chapter, we mentioned that NATS is designed to cater to the requirements of modern distributed systems while providing a simple interface to work with the clients. That is why the NATS protocol, which defines the interaction between the NATS server and the NATS client, has a simple set of commands that are based on plain text messages with a publish and subscribe type model. A simple TCP client such as Telnet can communicate with the NATS server using these protocol commands.

Protocol conventions

The NATS protocol uses a set of conventions to make sure that both the client and server understand the messages that are shared:

- **Control line with optional content**: Every communication that happens between the client and the NATS server consists of a command message followed by optional data. Only the PUB and MSG commands carry a payload (data) with the command.

- **Field delimiters**: The different parameters and options in the command are separated by whitespace or a tab. Consecutive whitespace characters in the command are considered as single delimiting characters.

- **Newlines**: NATS uses a newline character sequence (**Carriage Return (CR)** and **Line Feed (LF)**) to mark the end of the protocol message, as well as the message payload in a PUB or MSG protocol message.

- **Subject names**: The names of the subjects that are used when publishing and subscribing messages are case-sensitive and must contain alphanumeric characters with no whitespaces and cannot be empty. Subject names cannot contain the . and > special characters since those have a special purpose within the NATS messaging protocol. Hierarchical subject names separated by (.) are also allowed. A few sample subject names are mentioned here:

  - `PATIENT`, `NAME`, `patient.name`, `patient.NAME`, `PATIENT.NAME`, and `PATIENT.NAME.LAST` are all valid subject names.

  - `patient. name`, `patient..name`, and `patient..` are not valid subject names.

  - Subject names consist of one or more tokens separated by a dot (.) character. The > character can only be used as the last token of a subject name and it matches all the tokens after that. The * character matches a single token in the position it was listed.

- **Character encoding**: The NATS server supports ASCII character encoding for subject names and clients should adhere to the same as a best practice. There is no guarantee of support for non-ASCII encodings such as UTF-8 for subject names, even though it may be supported in certain cases.

With these conventions in place, the NATS protocol defines a set of commands or messages to communicate between the client and the server.

Protocol messages

NATS only has 10 simple protocol messages to communicate between the client and the server. Those commands are mentioned in the following table:

Command Name	Sent By	Description
INFO	Server	The server sends this message to the client after the initial request from the client.
CONNECT	Client	Once a connection has been established, the client uses this command to specify additional connection details.
PUB	Client	Send a message to the server with a subject name and an optional reply subject name.
SUB	Client	Subscribe to the NATS server on a given subject.
UNSUB	Client	Remove interest in a given subject and stop receiving messages.
MSG	Server	Use to deliver the message to the interested subscribers.
PING	Both	This message is used to keep the connection between a client and the server alive. When one party sends this message, the other party should respond with a PONG message to make a successful keep-alive connection.
PONG	Both	This is the response message to the PING message that is used to keep the connection alive.
+OK	Server	This is used by the server to let the client know that the previous message that was sent by the client was a well-formatted message. This is only sent in verbose mode.
-ERR	Server	This is used by the server to let the client know that there has been a protocol error.

One of the benefits of using a plaintext protocol is that it also makes it easier to interact with the server manually with a simple client such as Telnet, which all of us have in our working environments. Let's go through each of the aforementioned commands with an example to understand the protocol in detail.

Setting up the environment

NATS provides multiple options to install the NATS server in different environments. We'll follow the most generic approach in which we will download the binary from GitHub and install it. The latest version of the NATS server, at the time of writing, is 2.2.2 and it is available at `https://github.com/nats-io/nats-server/releases/tag/v2.2.2`.

Once the server has been installed, you can start the server with the following command:

Listing 3.1 – Starting the NATS server

```
$ nats-server
[180374] 2021/04/23 17:19:18.767627 [INF]  Starting nats-server
[180374] 2021/04/23 17:19:18.767830 [INF]     Version:   2.2.2
[180374] 2021/04/23 17:19:18.767861 [INF]     Git:       [a5f3aab]
[180374] 2021/04/23 17:19:18.767882 [INF]
   [Name:          NDPPCEM4HMLT5KMCDRQRTLKVPFEULBLFWCJLQ36
   [X5BYWFGWXQZRR6EYS
[180374] 2021/04/23 17:19:18.767902 [INF]     ID:
   NDPPCEM4HMLT5KMCDRQRTLKVPFEULBLFWCJLQ36X5BYWFGWXQZRR6EYS
[180374] 2021/04/23 17:19:18.771686 [INF] Listening for client
connections on 0.0.0.0:4222
[180374] 2021/04/23 17:19:18.774127 [INF] Server is ready
```

Connecting to the NATS server

Once the server has been started, we can connect to the server using a Telnet client:

Listing 3.2 – Connecting to the NATS server using a Telnet client

```
$ telnet 127.0.0.1 4222
Trying 127.0.0.1...
Connected to 127.0.0.1.
```

```
Escape character is '^]'.
INFO {"server_id":"NDPPCEM4HMLT5KMCDRQRTLKVPFEULBLFWCJ
LQ36X5BYWFGWXQZRR6EYS","server_name":
"NDPPCEM4HMLT5KMCDRQRTLKVPFEULBLFWCJLQ36X5BYWFGWXQZRR6EYS",
"version":"2.2.2","proto":1,"git_commit":"a5f3aab","go":
"go1.16.3","host":"0.0.0.0","port":4222,"headers":true,"max_
payload":1048576,"client_id":3,"client_ip":"127.0.0.1"}
```

In this scenario, the server responds with the INFO message containing a JSON string with information about how to handle the connection to the server. It contains details about the following:

- Server ID
- Server name
- Server version
- Port
- Maximum payload
- Client ID
- Client IP

Sending and receiving messages

Now that we are connected to the server, let's publish a message to the server using the PUB command. In the same terminal window that you executed the Telnet command in, you can execute the following command:

Listing 3.3 – Publishing a message to NATS

```
PUB patient.profile 5
Chris
+OK
```

You could observe the **+OK** message coming from the server as a response to the PUB command. This means that the server has accepted the message properly. It does not provide any details about whether the subscribers have received the message or not since they are fully decoupled.

Now, let's start two separate terminal windows and use one window as the publisher and the other window as the subscriber. From the subscriber window, we can connect to the server using the Telnet command (*Listing 3.1*) and then execute the following command:

Listing 3.4 – Subscribing to a subject in NATS

```
SUB patient.profile 90
+OK
```

Now, if we publish a message from the publisher terminal using the command mentioned in *Listing 3.3*, the subscriber will receive the message with the following entry in the terminal window:

Listing 3.5 – Receiving a message

```
MSG patient.profile 90 5
Chris
```

Here, the same message we published on the `patient.profile` topic is received by the subscriber.

Regarding the PUB and SUB commands, we observed that the server is sending back an acknowledgment in the form of an **+OK** message. This message is an indication of whether the particular operation (PUB/SUB) has been successfully executed at the server. We can turn this off by using the CONNECT command, as shown here:

Listing 3.6 – Updating the connection parameters

```
CONNECT {"verbose":false}
```

There can be situations where clients need to stop receiving messages from the server. Clients can use the UNSUB command to unsubscribe from a given subject to stop receiving messages, as shown here:

Listing 3.7 – Unsubscribing from a subject

```
UNSUB 90
```

In this command, we provide the subscription ID (90) that was used when subscribing to the subject in the SUB command. This command has an option to automatically unsubscribe after receiving a certain number of messages. The command to achieve auto unsubscribe is mentioned here:

Listing 3.8 – Unsubscribing from a subject after five messages

```
UNSUB 90 5
```

In addition to these commands, NATS uses two protocol messages called PING and PONG to keep the connection between the client and server alive. These commands help the server manage the resources efficiently by closing connections with idle clients. Here, you can see that the PING message is received by the client if you observe the terminal window that was used to create the connection for a couple of minutes:

Listing 3.9 – The PING command received from the server

```
$ telnet 127.0.0.1 4222
Trying 127.0.0.1...
Connected to 127.0.0.1.
Escape character is '^]'.
INFO {"server_id":"NAYAASEAFI4W4JFBHVEU6OQARS76M3RHJL2ZFNJRL2
GXJL2I4IUXQ6BD2","server_name":"NAYASEAFI4W4JFBHVEU6OQARS76M3
RHJL2ZFNJRL2GXJL2I4IUXQ6BD2","version":"2.2.2","proto":1,
"git_commit":"a5f3aab","go":"go1.16.3","host":"0.0.0.0",
"port":4222,"headers":true,"max_payload":1048576,"client_
id":5,"client_ip":"127.0.0.1"}
PING
```

Once this message (PING) has been received, the client should respond with a PONG message if there is no other activity going on. If the client does not respond with a PONG message for two continuous PING messages, the server will disconnect the client from the server. The following code demonstrates a scenario where the server has automatically tiered down the connection due to a lack of activity from the client:

Listing 3.10 – The NATS server automatically disconnects the client due to a lack of activity

```
$ telnet 127.0.0.1 4222
Trying 127.0.0.1...
Connected to 127.0.0.1.
Escape character is '^]'.
INFO {"server__id":"NAYASEAFI4W4JFBHVEU6OQARS76M3RHJL2ZFNJR
L2GXJL2I4IUXQ6BD2","server_name":"NAYASEAFI4W4JFBHVEU6OQARS7
6M3RHJL2ZFNJRL2GXJL2I4IUXQ6BD2","version":"2.2.2","proto":
1,"git_commit":"a5f3aab","go":"go1.16.3","host":"0.0.0.0",
"port":4222,"headers":true,"max_payload":1048576,"client_
id":4,"client_ip":"127.0.0.1"}
pub patient.profile 5
hello
+OK
pub patient.profile 5
hello
+OK
PING
PING
-ERR 'Stale Connection'
Connection closed by foreign host.
```

In the preceding listing, we can see the ERR command, which is received by the client from the server with a message stating Stale Connection. This carries the details of the error that occurred.

Advanced NATS concepts

Using NATS in production-grade systems requires advanced capabilities such as clustering, monitoring, and security. We will cover these advanced concepts in this section.

Clustering the NATS server

NATS is designed to support complex distributed systems at a global scale. NATS supports clustering to deploy multiple NATS-server nodes connected in a full mesh topology to provide improved reliability and failure handling. If a server goes down for some reason, the client can automatically connect to a different server that contains a replica of the message. NATS servers form a mesh dynamically by gossiping and connecting to all the known servers without needing to preconfigure the nodes. Hence, clusters can grow, shrink, and self-heal.

Once the cluster has been set up, the NATS servers replicate messages by forwarding the messages to connected servers. It has a forwarding limit of 1 node. Every server on the cluster forwards the messages to one of the adjacent servers that has a connection to it. Once such a message is received by an adjacent server, that message can only be sent to the clients connected to that server. The following diagram depicts a scenario where a client publishes a message to one particular node (nats-server-1) and the subscriber receives the message from a different node (nats-server-3) that is adjacent to the first node:

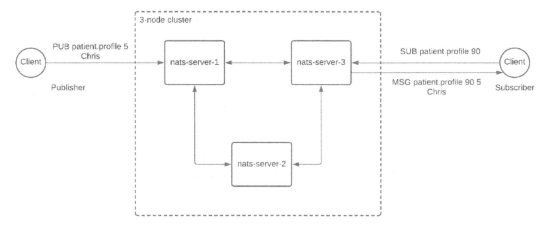

Figure 3.8 – NATS server cluster

As shown in the preceding diagram, the client (publisher) sends a message to the `patient.profile` subject to the NATS server cluster by connecting to nats-server-1. Then, this message is forwarded to nats-server-3 and the client (subscriber) subscribes to this particular node (nats-server-3) on the same subject name. Since the message is forwarded to this node (nats-server-3), the client (subscriber) receives the message.

Let's assume a practical scenario where the nats-server-1 node goes down due to a hardware failure. Here, the cluster should still be able to operate without any failure. The following diagram depicts a similar scenario:

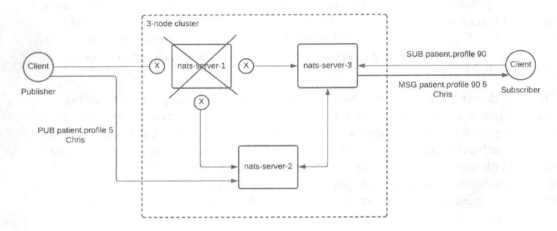

Figure 3.9 – NATS cluster node failure

As shown in the preceding diagram, once the nats-server-1 node goes down, the client will automatically identify the node failure and send the message to the next available node, which is nats-server-2. Then, it will forward the message to nats-server-3 since that is the only adjacent node available. Once the subscriber client subscribes to the same node (nats-server-3), it will receive the message without any failure.

The NATS server has really good performance when it comes to handling large volumes of messages. Hence, in most cases, a three-server cluster would be sufficient until you need to handle billions of messages per day. One thing to note here is that since NATS server clusters form a full-mesh topology, adding each node introduces more connections to the cluster and hence more TCP traffic over the network. As an example, a cluster of 5 nodes has 10 connections, while a cluster of 10 nodes has 45 connections. Hence, understanding the capacity requirements of the system and using the required number of nodes optimally is crucial.

Configuring a NATS server cluster

Let's try to set up a NATS server cluster with three servers running locally. In a real-world scenario, these three servers would run on three separate nodes. When starting the NATS server in cluster mode, we need to specify the following parameters:

- `--cluster`: This specifies the URL that listens to cluster connections from other servers.

- `--routes`: This specifies the URL that connects to an external cluster seed server.

- `--cluster_name`: This will set the name of the cluster so that servers can connect to that same cluster. If this parameter is not provided, every time we start a server, it will generate a random cluster name and update the name across all the cluster nodes:

Listing 3.11 – Starting the first server (seed server) of the cluster

```
$ nats-server -p 4222 -cluster nats://localhost:4248 --cluster_
name test_cluster
[209466] 2021/04/25 11:42:46.662878 [INF] Starting nats-server
[209466] 2021/04/25 11:42:46.663179 [INF]    Version:   2.2.2
[209466] 2021/04/25 11:42:46.663216 [INF]    Git:        [a5f3aab]
[209466] 2021/04/25 11:42:46.663246 [INF]    Name:
NCJBD7YI4EEKUH4XL5C7CQ3MKF6JI7RE7FCRUMQ5U25ECCY3BCV6FLZY
[209466] 2021/04/25 11:42:46.663274 [INF]    ID:
NCJBD7YI4EEKUH4XL5C7CQ3MKF6JI7RE7FCRUMQ5U25ECCY3BCV6FLZY
[209466] 2021/04/25 11:42:46.665870 [INF] Listening for client
connections on 0.0.0.0:4222
[209466] 2021/04/25 11:42:46.668852 [INF] Server is ready
[209466] 2021/04/25 11:42:46.672936 [INF] Cluster name is test_
cluster
[209466] 2021/04/25 11:42:46.675480 [INF] Listening for route
connections on localhost:4248
```

As mentioned in the preceding code, the first NATS server is started as the seed server that other servers will connect to using the cluster URL (`nats://localhost:4248`) that's been published here. Now, we can start the second server with the following command:

Listing 3.12 – Starting the second server of the cluster

```
$ nats-server -p 5222 -cluster nats://localhost:5248 -routes
nats://localhost:4248 --cluster_name test_cluster
[209501] 2021/04/25 11:43:18.196749 [INF] Starting nats-server
[209501] 2021/04/25 11:43:18.196977 [INF]    Version:  2.2.2
[209501] 2021/04/25 11:43:18.197015 [INF]    Git:      [a5f3aab]
[209501] 2021/04/25 11:43:18.197046 [INF]    Name:
ND7BUNN2GOVCEFDMOGHOHXQFDOVZRLTW5B4MGA7LYBAHE4FMEWQCOT4A
[209501] 2021/04/25 11:43:18.197077 [INF]    ID:
ND7BUNN2GOVCEFDMOGHOHXQFDOVZRLTW5B4MGA7LYBAHE4FMEWQCOT4A
[209501] 2021/04/25 11:43:18.206064 [INF] Listening for client
connections on 0.0.0.0:5222
[209501] 2021/04/25 11:43:18.209149 [INF] Server is ready
[209501] 2021/04/25 11:43:18.209205 [INF] Cluster name is test_
cluster
[209501] 2021/04/25 11:43:18.211080 [INF] Listening for route
connections on localhost:5248
[209501] 2021/04/25 11:43:18.217194 [INF] 127.0.0.1:4248 -
rid:3 - Route connection created
```

Once the second server has been started with the `-routes` option, it will create a connection to the first node over that URL. At this time, you could observe that the first node also prints a log entry containing the connection's creation, as shown here:

Listing 3.13 – The first server connects to the second server

```
[209466] 2021/04/25 11:43:18.215990 [INF] 127.0.0.1:33878 -
rid:3 - Route connection created
```

Now that we have two servers running, let's start the third server using the following command:

Listing 3.14 – Starting the third server of the cluster

```
$ nats-server -p 6222 -cluster nats://localhost:6248 -routes
nats://localhost:4248 --cluster_name test_cluster
[209531] 2021/04/25 11:43:45.926996 [INF] Starting nats-server
[209531] 2021/04/25 11:43:45.927255 [INF]   Version:   2.2.2
[209531] 2021/04/25 11:43:45.927274 [INF]   Git:       [a5f3aab]
[209531] 2021/04/25 11:43:45.927324 [INF]   Name:
NA2OE2UVEQNDCYFKCKWPYXLNM7M4WQVXN7QWP2TF7BGDJLMJBDIER334
[209531] 2021/04/25 11:43:45.927356 [INF]   ID:
NA2OE2UVEQNDCYFKCKWPYXLNM7M4WQVXN7QWP2TF7BGDJLMJBDIER334
[209531] 2021/04/25 11:43:45.936813 [INF] Listening for client
connections on 0.0.0.0:6222
[209531] 2021/04/25 11:43:45.940094 [INF] Server is ready
[209531] 2021/04/25 11:43:45.940163 [INF] Cluster name is test_
cluster
[209531] 2021/04/25 11:43:45.942566 [INF] Listening for route
connections on localhost:6248
[209531] 2021/04/25 11:43:45.947403 [INF] 127.0.0.1:4248 -
rid:3 - Route connection created
[209531] 2021/04/25 11:43:45.956018 [INF] 127.0.0.1:45748 -
rid:4 - Route connection created
```

Here, again, we specify the seed server using the -routes option when starting the server by specifying the URL. Here, we can see that this third server has created two route connections to the first two servers. At the same time, we can see the same operation resulting in additional log entries in the first and second servers, as shown in the following code examples:

Listing 3.15 – The first server creating a route connection to the third server

```
[209466] 2021/04/25 11:43:45.946299 [INF] 127.0.0.1:33892 -
rid:4 - Route connection created
```

Listing 3.16 – The second server creating a route connection to the third server

```
[209501] 2021/04/25 11:43:45.956062 [INF] 127.0.0.1:6248 -
rid:4 - Route connection created
```

At this point, we have the three-server cluster up and running in fully connected mode. Let's try and publish a message and subscribe to it from a different server to validate the concepts we discussed in the previous section.

Let's create a connection to the first server by using the Telnet client and subscribe to the patient.profile subject in the same server using the following commands:

Listing 3.17 – Connecting to the first server and subscribing to a subject

```
$ telnet localhost 4222
Trying 127.0.0.1...
Connected to localhost.
Escape character is '^]'.
INFO {"server_ id":"NCJBD7YI4EEKUH4XL5C7CQ3MKF6JI7RE7FCRUM
Q5U25ECCY3BCV6FLZY","server_name":"NCJBD7YI4EEKUH4XL5C7
CQ3MKF6JI7RE7FCRUMQ5U25ECCY3BCV6FLZY","version":"2.2.2",
"proto":1,"git_commit":"a5f3aab","go":"go1.16.3",
"host":"0.0.0.0","port":4222,"headers":true,"max_
payload":1048576,"client_id":5,"client_
ip":"127.0.0.1","cluster":"test_cluster","connect_urls":["192.1
68.1.7:4222","192.168.122.1:4222","172.19.0.1:4222","172.17.0.1
:4222","172.20.0.1:4222","172.18.0.1:4222","192.168.1.7:5222","
192.168.122.1:5222","172.19.0.1:5222","192.168.122.1:6222","172
.17.0.1:5222","172.20.0.1:5222","172.18.0.1:5222","192.168.1.7:
6222","172.19.0.1:6222","172.17.0.1:6222","172.20.0.1:6222","17
2.18.0.1:6222"]}
sub patient.profile 90
+OK
```

Now, we have one client connected to the first server, who has subscribed to the patient.profile subject. Let's create another client connection to the third server and publish a message using the same subject mentioned in the following code:

Listing 3.18 – Connecting to the third server and publishing a message

```
$ telnet localhost 6222
Trying 127.0.0.1...
Connected to localhost.
Escape character is '^]'.
INFO {"server_ id":"NA2OE2UVEQNDCYFKCKWPYXLNM7M4WQVXN7
QWP2TF7BGDJLMJBDIER334","server_
name":"NA2OE2UVEQNDCYFKCKWPYXLNM7M4WQVXN7QWP2TF7BGDJ
LMJBDIER334","version":"2.2.2","proto":1,"git_commit":
"a5f3aab","go":"go1.16.3","host":"0.0.0.0",
"port":6222,"headers":true,"max_payload":1048576,"client_
id":5,"client_ip":"127.0.0.1","cluster":"test_
cluster","connect_urls":["192.168.1.7:6222","192.168.122.1:622
2","172.19.0.1:6222","172.17.0.1:6222","172.20.0.1:6222","172.
18.0.1:6222","172.20.0.1:4222","172.17.0.1:5222","172.20.0.1:5
222","192.168.122.1:4222","172.19.0.1:4222","172.17.0.1:4222",
"192.168.122.1:5222","172.19.0.1:5222","172.18.0.1:5222","192.
168.1.7:4222","172.18.0.1:4222","192.168.1.7:5222"]}
pub patient.profile 5
Chris
+OK
```

With that, we have connected to the third server using port 6222 and published a message using the same subject; that is, patient.profile. Now, if we look at the first client terminal, which we used to connect to the first server and subscribe to the subject, we will see that the message has been delivered to that client. The log entry would look similar to the following:

Listing 3.19 – Message received through the first node

```
MSG patient.profile 90 5
Chris
```

The message is received by the client that is connected to the first server, even though we published the message to the third server. The clients can be designed in such a way that if there is a connection failure to a given server, it can reconnect to the next available server.

There is no explicit configuration for the seed server. It simply serves as the starting point for server discovery by other members of the cluster, as well as clients. As such, these are the servers that clients have in their list of connecting URLs and that cluster members have in their list of routes. They reduce configuration as not every server needs to be in these lists. But the ability for other servers and clients to successfully connect depends on the seed server that's running. If multiple seed servers are being used, they make use of the `routes` option as well so that they can establish routes to one another.

Monitoring NATS

In a production deployment, it is necessary to monitor the server's health and performance-related information to keep the system in good condition, as well as to make scaling decisions. The NATS server provides an HTTP server on a dedicated monitoring port to monitor the server. This monitoring server exposes several endpoints, such as the following:

- General server information
- Connections
- Routing
- Gateways
- Leaf nodes
- Subscription routing
- Account information

All these endpoints provide a JSON response and its details. The monitoring server can be enabled by providing the relevant flags at startup, as mentioned in the following code:

Listing 3.20 – Starting the NATS server with monitoring enabled

```
$ nats-server -m 8222
[215662] 2021/04/25 18:38:07.365515 [INF] Starting nats-server
[215662] 2021/04/25 18:38:07.365793 [INF]   Version:  2.2.2
[215662] 2021/04/25 18:38:07.365973 [INF]   Git:      [a5f3aab]
[215662] 2021/04/25 18:38:07.366087 [INF]   Name:
NBEKLGAUANEE7EW67KU7NPQ4X3N65FY4DFPKYHBSN7HA76QKK6KZFGP7
[215662] 2021/04/25 18:38:07.366117 [INF]   ID:
NBEKLGAUANEE7EW67KU7NPQ4X3N65FY4DFPKYHBSN7HA76QKK6KZFGP7
```

```
[215662] 2021/04/25 18:38:07.371691 [INF] Starting http monitor
 on 0.0.0.0:8222
[215662] 2021/04/25 18:38:07.371971 [INF] Listening for client
connections on 0.0.0.0:4222
[215662] 2021/04/25 18:38:07.375373 [INF] Server is ready
```

Once the server has been started, the monitoring server can be accessed via http://localhost:8222. Let's look at the metrics that are exposed by the monitoring server.

General server information

The /varz endpoint provides general server information such as server ID, uptime, number of connections, CPU and memory usage, ports, timeouts, and request counts, to name a few. It also provides statistics about server performance such as in/out messages and in/out bytes. This information is available here: http://localhost:8222/varz.

Connections

The /connz endpoint provides additional details on the current and recently closed connections. There are a set of query parameters available to filter this information based on certain conditions. The following are a few examples of requesting connection information from the server:

- Get up to 1,024 connections: http://demo.nats.io:8222/connz

- Get closed connection information: http://demo.nats.io:8222/connz?state=closed

Routing

The /routez endpoint provides details of the active routes for a cluster. This information is available at http://localhost:8222/routez.

Gateway

The /gatewayz endpoint reports information about the gateways that are used to create a NATS supercluster. This information is available at http://localhost:8222/gatewayz.

Leaf nodes

The /leafz endpoint provides details on the leaf node connections in a cluster. This information is available at http://localhost:8222/leafz.

Subscription routing

The `/subsz` endpoint reports detailed information about the current subscriptions and the routing data structure. This information is available at `http://localhost:8222/subsz`.

Account information

The `/accountz` endpoint reports information about a server's active accounts. The default behavior is to return a list of all accounts known to the server. This information is available at `http://localhost:8222/accountz`.

Security with NATS

NATS supports security at multiple levels to enable secure communication within the ecosystem. There are three main areas of security that can be configured in NATS:

- Secure communication with **Transport Layer Security (TLS)**
- Authentication of clients
- Authorization of clients

Let's discuss each area in detail.

Secure communication with TLS

The NATS server uses modern encryption methods for communicating using TLS. It can be configured for the following connection types:

- Client connections
- Route connections
- Monitoring connections

The server keeps the configurations related to TLS in a configuration map with a set of properties. These properties contain details such as the certification file, key file, CA file, cipher suites, timeout, and other security-related parameters. These values can be set using a configuration file or via the startup parameters.

Once the server has been set up with TLS for communication, all the connections need to be established securely. The clients will start the connection in an insecure channel and once the server responds with the server details via the INFO message, the client starts the TLS handshake and upgrades the connection to a secure connection. It then uses that secure connection to continue the communication from that point on:

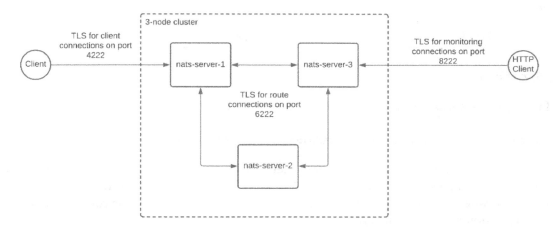

Figure 3.10 – Secure NATS cluster

The preceding diagram depicts how the NATS server can be configured with TLS-based security for clients, routes, and monitoring connections.

Authenticating clients

The NATS server supports different methods for authenticating client connections. These authentication methods include the following:

- Token authentication
- Basic authentication with a username/password
- Certificate-based authentication
- NKEY with a challenge
- JW-based authentication

Let's discuss each authentication method in detail.

Token authentication

This is the simplest authentication method in NATS. In this method, a string is configured on the server side in a configuration file or as a startup parameter that will be used by clients as the token when connecting with the NATS server. The following code snippets show how to start the server with a security token and how a client can connect to the server using the same token:

Listing 3.21 – Starting the NATS server with a token

```
$ nats-server --auth s3cr3t -p 5222
```

Listing 3.22 – Connecting to the NATS server with a token

```
$ nats sub -s nats://s3cr3t@localhost:5222 ">"
```

In the preceding code, the NATS subscriber client that is available in the NATS examples repository is being used to connect to the NATS server using the token that was configured when we started the server in *Listing 3.21*. This token can also be configured in a configuration file, as shown here:

Listing 3.23 – Configuring token security using a configuration file

```
authorization {
    token: "s3cr3t"
}
```

Tokens can be encrypted (bcrypted) by enabling an additional layer of security as the cleartext version of the token would not be persisted on the server configuration file.

Basic authentication with a username/password

The NATS server can also be protected with basic authentication using a similar approach. The username/password details can be stored in the configuration file, as shown here:

Listing 3.24 – Configuring the username/password security in NATS using the configuration file

```
authorization: {
users: [
{user: a, password: b},
{user: b, password: a}
```

```
    ]
}
```

These configurations can also be encrypted using the tools available in the NATS ecosystem.

Certificate-based authentication

The server can require TLS certificates from a client. When needed, you can use the certificates to do the following:

- Validate that the client certificate matches a known or trusted CA
- Extract information from a trusted certificate to provide authentication

The server can verify a client certificate using a CA certificate. To require verification, add the option `verify` to the TLS configuration section, as shown here:

Listing 3.25 – Configuring client certificate validation in the NATS server

```
tls {
    cert_file:  "server-cert.pem"
    key_file:   "server-key.pem"
    ca_file:    "rootCA.pem"
    verify:     true
}
```

This option verifies that the client's certificate has been signed by the CA specified in the `ca_file` option. When `ca_file` is not present, it will default to the CAs in the system trust store.

In addition to verifying that a specified CA issued a client certificate, you can use information encoded in the certificate to authenticate a client. The client wouldn't have to provide or track usernames or passwords. To have the TLS mutual authentication map certificate attributes to the user's identity, use `verify_and_map`, as shown here:

Listing 3.26 – Configuring client certificate validation with user mapping

```
tls {
    cert_file: "server-cert.pem"
    key_file:  "server-key.pem"
    ca_file:   "rootCA.pem"
    # Require a client certificate and map user id from
```

```
    certificate
  verify_and_map: true
}
```

When present, the server will check if a **Subject Alternative Name (SAN)** maps to a user. It will search all the email addresses first, then all DNS names. If no user is found, it will try the certificate subject.

NKEY with a challenge

NKEYs are a new, highly secure public-key signature system based on Ed25519: `https://ed25519.cr.yp.to/`. With NKEYs, the server can verify identities without ever storing or seeing private keys. The authentication system works by requiring a connecting client to provide its public key and digitally signs a challenge with its private key. The server generates a random challenge with every connection request, making it immune to playback attacks. The generated signature is validated against the provided public key, thus proving the identity of the client. If the public key is known to the server, authentication succeeds.

To use `nkey` authentication, add a user and set the `nkey` property to the public key of the user you want to authenticate, as shown here:

Listing 3.27 – Configuring NKEY-based security in the NATS server

```
authorization: {
  users: [
    { nkey: UDXU4RCSJNZOIQHZNWXHXORDPRTG
      NJAHAHFRGZNEEJCPQTT2M7NLCNF4 }
  ]
}
```

Note that the `users` section sets the `nkey` property (user/password/token properties are not needed). Add permission sections as required.

JWT-based authentication

JSON Web Token (**JWT**)-based authentication allows you to configure distributed security within a NATS deployment. With other authentication mechanisms, the configuration for identifying a user and account is in the server configuration file. JWT authentication leverages (JWTs) to describe the various entities that are supported. When a client connects, servers verify the authenticity of the request using NKEYs, download account information, and validate a trust chain. Users are not directly tracked by the server, but rather verified as belonging to an account. This allows users to be managed without the need for server configuration updates.

Authorizing clients

The NATS server supports authorization using subject-level permissions on a per-user basis. Permission-based authorization is available with multi-user authentication via the user's list. Each permission specifies the subjects the user can publish to and subscribe to. For a more complex configuration, you can specify a permission object that explicitly allows or denies subjects. The specified subjects can specify wildcards as well. Permissions can make use of variables. These permissions can be configured with a configuration file, as shown here:

Listing 3.28 – Configuring authorization of clients using a config file

```
authorization: {
users = [
{
user: admin
password: secret
permissions: {
publish: ">"
subscribe: ">"
}
}
{
user: test
password: test
permissions: {
publish: {
deny: ">"
},
```

```
subscribe: {
allow: "client.>"
  }
 }
 }
]
}
```

Let's take a look at this configuration:

- The `admin` user can publish to any subject and subscribe to any subject.
- The `test` user cannot publish to any subject, but it can listen to subjects starting with `client`.

NATS provides a comprehensive security model where permissions can be configured for advanced use cases with variables.

Advantages of NATS

So far, we have discussed the features of the NATS messaging system and how these features can be configured with a few examples. In this section, we'll discuss the advantages of NATS in general and its advantages when comparing it with other popular messaging platforms. The following is a set of advantages of NATS as a messaging system (or message broker) in general:

- **High performance**: NATS has the best performance compared to many other brokers, such as Kafka, RabbitMQ, and ActiveMQ. It can process millions of messages per second within a single server with moderate hardware. It can reduce hardware costs as well as network and management costs.
- **Resilient**: NATS provides automatic scaling and self-healing capabilities to build a resilient messaging platform for users by hiding internal node failures. Also, it allows you to scale the servers up and down according to the load without letting the clients know about these changes.
- **Secure**: NATS provides security at multiple levels to enable end-to-end security for data, users, and systems without compromising the simplicity of the platform. With true multi-tenancy, you can securely isolate and share your data to fully meet your business needs, mitigating risk and achieving faster time to value.

- **Agile**: NATS deployments can be started with a simple three-node cluster and scale up with simple configurations in an agile manner, without the need for complex clustering configurations and deployment models.

In addition to these generic advantages, NATS is often compared to other messaging technologies to understand the value it brings to enterprise platforms. The official NATS website provides a comparison of NATS with other popular messaging technologies, such as Kafka, RabbitMQ, gRPC, and Pulsar, which you can find at `https://docs.nats.io/compare-nats`.

Summary

In this chapter, we discussed the concepts of the NATS messaging platform by going through the NATS protocol, messaging models, clustering, monitoring, and security with a few practical examples. We also covered some of the advantages of NATS and how it compares to other messaging platforms, such as Kafka, RabbitMQ, and gRPC. This knowledge helps developers and architects understand the NATS messaging technology concepts and how it works in the real world so that they can build effective enterprise platforms while using NATS messaging. This chapter concludes the first section of this book, where we discussed the basic concepts of the microservice architecture, messaging technologies, and the NATS technology.

In the next section, we will discuss how to use this knowledge to design a microservice architecture alongside the NATS messaging platform. In the next chapter, we will discuss how to use the NATS messaging platform in a microservice architecture for inter-service communication.

Further reading

To learn more about what was covered in this chapter, please take a look at the following resources:

- *NATS documentation*: `https://docs.nats.io/`.
- *Practical NATS: From beginner to pro*, by Waldemar Quevedo, available at `https://www.oreilly.com/library/view/practical-nats-from/9781484235706/`.
- *NATS client tools*: `https://github.com/nats-io/nats.go`.

Section 2: Building Microservices with NATS

This section describes how NATS fits into the microservices architecture. It provides a reference for building an enterprise software platform using the microservices architecture and NATS messaging. It also provides a reference implementation with an example application written in Go. This section also provides details of how to implement security and observability within a microservices architecture that has NATS messaging for inter-service communication.

This section contains the following chapters:

- *Chapter 4, How to Use NATS in a Microservices Architecture*
- *Chapter 5, Designing a Microservices Architecture with NATS*
- *Chapter 6, A Practical Example of Microservices with NATS*
- *Chapter 7, Securing a Microservices Architecture with NATS*
- *Chapter 8, Observability with NATS in a Microservices Architecture*

4
How to Use NATS in a Microservice Architecture

In the first part of this book, we discussed the evolution of distributed systems, the microservice architecture, messaging technologies, and NATS messaging in detail as separate topics. In this second part of this book, we will start by using what we learned previously to design an effective microservice architecture by exploring these topics further and the correlation among them in detail.

As we discussed in *Chapter 2, Why Is Messaging Important in the Microservice Architecture?*, microservice architecture-based platforms have two main message flows called **North-South traffic** and **East-West traffic**. Sometimes, East-West traffic is also called **inter-service communication** in a microservice architecture. In this chapter, we will discuss these messaging flows in detail and how NATS can be used for inter-service communication within the microservice architecture. We are going to cover the following main topics in this chapter:

- Understanding the challenges of the microservice architecture
- What is inter-service communication and why is it important?

- How does NATS solve the inter-service communication challenge?
- Advantages of using NATS for inter-service communication

By the end of this chapter, you will understand the importance, challenges, and the NATS solution for inter-service communication in a microservice architecture.

Technical requirements

We will be using some code examples to demonstrate the practical usage of NATS for inter-service communication in a microservice architecture. Ensure you meet the following requirements in your development environment to try out the examples in this chapter:

- Install the latest JDK to run the Java examples
- Install Python3 to run the Python examples
- Install the Go programming language to run the Go examples
- Install the NATS server to test the client examples

The source code for the examples in this chapter can be found in the following GitHub repository: `https://github.com/PacktPublishing/Designing-Microservices-Platforms-with-NATS/tree/main/chapter4`.

Understanding the challenges of the microservice architecture

The microservice architecture offers many advantages such as agility, availability, and efficiency, to name a few. These advantages come with a set of challenges that need to be handled properly to gain the maximum output from the microservice architecture. Most of these challenges are related to the technical aspect of the solution, while there are also a few challenges related to organizational and team structure. Let's look at these challenges and understand what they are:

- Identifying the service boundaries
- Inter-service communication
- Securing services
- Monitoring services
- Organizational structures

Let's discuss each challenge in detail so that we can correlate these points when designing the solution with NATS.

Identifying the service boundaries

Microservices are great when you have them. But getting started with microservices is the most challenging part since you need to identify the boundaries and scopes of each microservice. There are many theories around this such as **domain-driven design (DDD)** and **function per service**. If we consider the OPD application that we discussed in *Chapter 1*, *Introduction to the Microservice Architecture*, we can define these services based on the functions within an OPD. Some examples are as follows:

- Patient registration
- Patient inspection
- Temporary treatment
- Discharge

To identify these functional areas or domains, we must have a proper understanding of how an OPD works within a hospital. This knowledge can be gained through various practical experiences and experiments. This means that, initially, we could assume that an OPD is just a room where patients come in, get some medicine, and go home or are admitted to the respective ward. But with time, as the number of patients who come into the hospital increases, the administration team might have to think about dividing the functions to separate units and manage those functions separately. That is where these individual functions and their scopes are defined and assigned to separate personnel. We also consider this approach as DDD since each function represents a particular domain or scope. There are a couple of key aspects that we should consider when defining these boundaries. These are **loose coupling** and **cohesion**.

Loose coupling allows the microservice architecture to be tolerant to change. This means that making a change to a given service or multiple services will not require making a change to other services. Each service can evolve its business logic internally if it adheres to the interface definitions without tightly coupling to any other service. The concept of cohesion is related to how well these individual services interact with each other to achieve the overall goal of the application. This means that when designing microservices based on domain, scope, or function, it is essential to make sure that they interact with each other whenever necessary to achieve this common goal.

What this shows is that it is always good to have domain experts and business people when designing a microservice architecture for a particular application. They could share their practical experiences so that we can utilize them to define the boundaries of the services that we are designing and developing.

Inter-service communication

Once the service boundaries have been identified, the services can be developed with the best possible technology that is available. This is where the concept of **polyglot programming** comes into the picture, where different services are implemented with different programming languages. It is fine to go with a polyglot programming model if we have a mechanism to integrate these services through inter-service communication. This is the main topic of this book, and we will cover this topic in detail throughout this book. The main challenge with inter-service communication is connectivity.

As we discussed in *Chapter 2, Why Is Messaging Important in the Microservice Architecture?*, point-to-point communication works well for small-scale systems. However, when we have more services, it becomes unmanageable. Hence, we looked at different options with Kafka, NATS, and gRPC, to name a few. In this chapter, we will be discussing how to use NATS for inter-service communication. When different services are utilizing different technologies, each of these services should be capable of communicating with each other using their native technologies or programming languages. That is the task of the NATS server, which acts as a centralized broker that can be accessed via the respective native technologies by utilizing the client SDKs that are available for NATS. We will discuss this in detail later in this chapter.

Securing services

There are two main concepts around securing services: **authentication** and **authorization**. The former relates to identifying the user and verifying the user based on the attributes they present while accessing the service or logging in. The latter relates to the permissions this authenticated user must have to perform certain tasks within the application. Microservices are more often developed to work with critical business data in the form of **Create**, **Read**, **Update**, and **Delete** (**CRUD**) operations. These services are accessed by distinct types of consumer applications. We can categorize these consumer applications as follows:

- Internal applications (used by employees)
- External applications (used by customers)
- Partner applications (used by partners)

When these applications access data that resides in an enterprise system, there needs to be security at two layers:

- **Transport Layer Security** (**TLS**) for encrypting data
- **Message Layer Security** (**MLS**) for controlling access to data

Nowadays, it is the general practice that TLS should be enabled for any traffic that goes through the internet. This is provided by the TLS and **Secure Socket Layer** (**SSL**) configurations. These are designed to provide communications security over a computer network via encryption. TLS 1.2 and TLS 1.3 are the versions that are recommended for communication at the time of writing. The following diagram shows how TLS can be enabled within a microservice architecture-based enterprise platform from the consumer application level to the internal applications that hold business data. We call this **end-to-end data security** or **security in transit**:

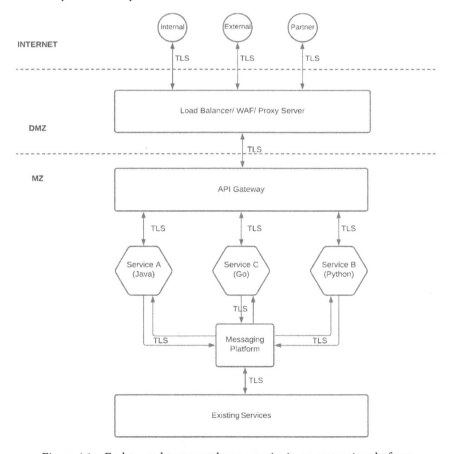

Figure 4.1 – End-to-end transport layer security in an enterprise platform

The preceding diagram shows end-to-end data encryption-based security being applied in an enterprise platform. Each communication link is enabled with TLS for data encryption. This is the maximum security level that can be achieved in terms of data encryption. However, this advanced security comes at a cost. This TLS-based communication adds a delay to communication. Having multiple TLS delays in end user communication can result in decreased performance at the user experience level. Hence, it is a decision that needs to be made after considering the use case and the expected experiences of the users.

Not all applications and systems require this level of security. If a certain application is more focused on performance and user experience is a critical factor of the application, we can come up with a solution that can save both the security aspect and the user experience aspect. The following diagram depicts such an approach where TLS is terminated at the API gateway layer and internal communications happen within the **Militarized Zone (MZ)** without it:

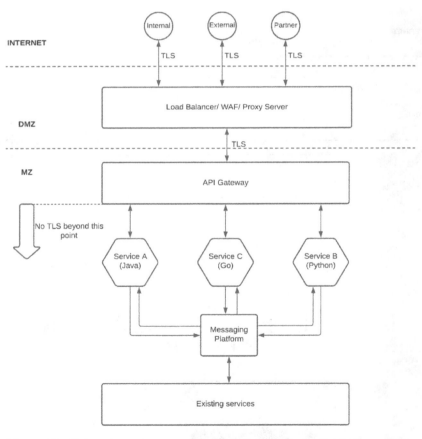

Figure 4.2 – Enterprise system secured with TLS with termination at the gateway

As per the preceding diagram, TLS is terminated at the API gateway layer and communications with microservices and beyond are done without TLS to save latency. This approach is used in many enterprises since internal system access can be protected with additional network-level security, where users cannot tap into the communication paths within the internal network. We'll discuss this in detail in *Chapter 7, SecuritySecuring a Microservices Architecture with NATS*.

The other aspect of security is the message-level security that is provided through the access control mechanisms implemented at each layer. These different consumer applications will have various levels of access that need to be controlled via mechanisms such as fine-grained authorization and role-based authentication, to name a few. In a monolithic application, the application itself handles security and it is a one-time task. But in a microservice architecture, there are hundreds of different services that need to secure the data that is accessed over the service, and implementing security at each service layer would be a tedious task. One possible solution for message-level security is using a **JSON Web Token (JWT)** to provide access to the services. This is depicted in the following diagram:

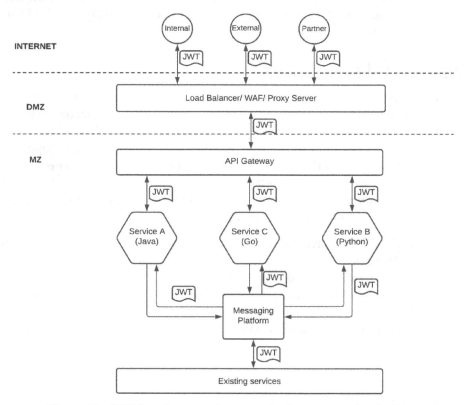

Figure 4.3 – JWT-based message layer security for an enterprise platform

As depicted in the preceding diagram, JWT is used across the board to authenticate and authorize the user's request. To implement this type of solution, all the components within the platform, including microservices, the messaging platform, and API gateways, should support JWT-based security. We will discuss this topic in more detail in *Chapter 7*, *Securing a Microservices Architecture with NATS*.

Monitoring services

We use monitoring tools to troubleshoot an issue when something goes wrong in the system. Before microservices, we only had to worry about a few services (monoliths), and going through the log files of those services and identifying the issue would have been simpler if the applications were written well with proper error logs. But with microservices, the number of services has increased, and that increases the number of log files and places that we must scrape through when identifying a root cause for an issue. Unless you have a proper monitoring solution that captures these log files from the services that generate a distributed view of the failure points, it will be a much harder task to troubleshoot an issue in a microservice architecture. We will discuss monitoring best practices and how to address these challenges in *Chapter 8*, *Observability with NATS in a Microservice Architecture*.

Organizational structures

In addition to the technical challenges that we mentioned in the previous sections, there are also non-technical challenges related to the microservice architecture. Most organizations used to work as separate teams with the **Center of Excellence** (**CoE**) that worked in siloed mode. This also leads to the waterfall-type software development methodologies, with a lot of friction and dependency between these CoE teams. According to Conway's law, "*any organization that designs a system (defined more broadly here than just information systems) will inevitably produce a design whose structure is a copy of the organization's communication structure.*"

This statement emphasizes the fact that if an organization has a hierarchical structure with siloed teams and a lot of dependencies between these teams, the software they develop will also be fragile and tightly coupled. The microservice architecture suggests an agile development approach with small teams (two pizza teams) that consist of different skilled members such as developers, architects, testers, designers, and DevOps, to name a few. We will discuss this topic in more detail in *Chapter 5*, *Designing a Microservice Architecture*.

With that, we have discussed the challenges of the microservice architecture at a higher level. We will explore these points in greater detail in this chapter as well as the upcoming chapters, as mentioned in the preceding section. We will focus more on the inter-service communication aspect and how NATS can be used to address the challenges of the same within this chapter in the upcoming sections. The next section will discuss inter-service communication and how NATS can play a role in successfully implementing it in a microservice architecture.

What is inter-service communication and why is it important?

In layman's terms, inter-service communication means sharing information between two applications (services) over a network in a distributed system. Sometimes, we call this **process messaging**. In *Chapter 2, Why Is Messaging Important in the Microservice Architecture?*, we discussed different messaging technologies and had a brief overview of how those technologies can be used in a microservice architecture. In this section, we will explore this topic in detail. The following diagram depicts the two types of message flows within a microservice architecture:

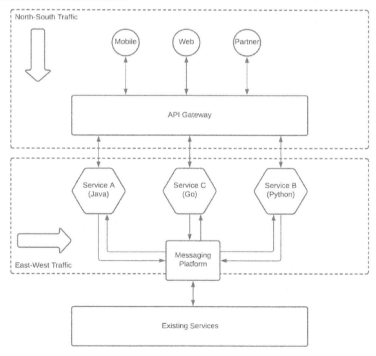

Figure 4.4 – Traffic flows within a microservice architecture

North-South traffic

The requests that come from applications and services outside the microservice platform toward the microservice platform via a service gateway such as an API gateway are considered North-South traffic.

East-West traffic

The requests flowing from one microservice to another microservice or a set of other microservices are called East-West traffic. This is also called inter-service communication in the context of microservices.

The microservice architecture offers development teams the freedom to select a technology stack and different programming languages for their respective services. This can lead to several challenges that were not there in the traditional monolithic development model, where all the components were implemented using the same technology and programming language. Here is a list of challenges that need to be addressed in inter-service communication within a polyglot-type microservice architecture:

- Selecting a common message format
- Defining service interfaces
- Accessing shared data
- Synchronous communication (request-reply) or asynchronous communication (event-based)

Let's discuss these points in detail and identify the best practices to address these when designing microservices.

Selecting a common message format

One principle that we must follow when designing complex systems is that we should make them as simple as possible. This requires some level of adherence to common standards while utilizing freedom. Different applications may use different message formats such as **eXtensible Markup Language (XML)**, **JavaScript Object Notation (JSON)**, plaintext, or binary, to name a few. Different messaging protocols such as HTTP provide mechanisms such as headers to specify the message format so that the recipient can process the message accordingly. But using different messaging formats within a microservice architecture for inter-service communication can cause unnecessary performance issues when converting across formats, such as JSON into XML. It would also cause data loss issues during the conversion. Due to these reasons, adhering to a common messaging format for inter-service communication makes communication simple and avoids issues such as performance degradation and data loss.

Defining service interfaces

Another key aspect of inter-service communication is the interface definition of the services. This definition would help other services understand the functionality provided by that service. There are a few important things to consider when selecting an interface definition method for services. These are as follows:

- The interface definition is simple to consume
- It hides implementation details
- It allows loose coupling between services
- It provides location independence

Services are developed to be consumed by clients. Hence, the interface should be clear and concise enough so that consumers can easily connect with it. At the same time, these interfaces should follow a common standard so that clients can utilize already developed SDKs and libraries for rapid development.

Given that microservices can follow a polyglot programming model, the interface should not reflect the underlying technology and should not expect the consumers to use the same technology. There are several standard interface definition languages and models that can be used to hide the implementation details of the service. A few examples are as follows:

- **Web Services Definition Language (WSDL)**
- **Open API Specification (OAS)**
- **Swagger**
- **Asynchronous APIs (AsyncAPI)**
- **Protocol Buffers (Protobuf)**

Each of these interface definition models represents a particular messaging model so that clients can consume the service without worrying about the underlying technology stack used to implement the service.

Accessing shared data

In a monolithic architecture, the application controls data access via in-memory techniques such as locks and semaphores, and it is guaranteed that no two components access data at the same time. But in a microservice architecture, these components become services running on independent runtimes without any coordination. As an example, if two services access the same file location, then there can be challenges such as one service reading the content before the other service completes writing the data, and partial data reads can cause applications to produce incorrect results. The microservice architecture suggests using local data stores to keep the service-related data while running separate microservices for data access, rather than sharing data over multi-access resources such as files or databases.

Synchronous or asynchronous communication

Different applications have different requirements in terms of how they process data. Depending on how soon the data needs to be processed, there are distinct types of approaches applications can follow, such as the following:

- Real time
- Near real time
- Batch

Most web-based applications need the results immediately. For example, if we are searching for a mobile phone on an e-commerce website, results need to appear as quickly as possible. This is a real-time data processing application that requires a synchronous communication model.

Let's extend the same use case to the next phase, where the user wants to purchase a mobile phone from the website. In this case, the request will be processed and fulfilled in near real time or in batch mode since there are multiple steps involved in shipping the mobile phone to the user from the vendor. This can follow an asynchronous communication approach where inter-service communication happens via a message broker.

The same application should process the customer's order and update the inventory systems at the end of each day. This can be done as a batch process asynchronously.

It is the responsibility of the architects to identify the right messaging models for various parts of the application. In a microservice architecture, both synchronous and asynchronous communication models need to be supported for inter-service communication. Hence, the message broker should be capable of handling both types of approaches. In this section, we identified the practical challenges of the microservice architecture and the challenges related to inter-service communication. Next, we will discuss how these challenges are addressed by the NATS messaging framework.

How NATS solves the inter-service communication challenge

Most of the great innovations are created to solve real-world problems and the more problems and challenges they solve and address, the better the solutions are, and they will stick around for a long time. NATS was created to solve the inter-service communication requirement within the Cloud Foundry platform. Cloud Foundry was a highly sophisticated application deployment and management platform that had many components that needed to communicate with each other in a highly efficient manner with greater scalability. NATS acted as the always-on dial tone for Cloud Foundry services. This means that the connecting services can consider the NATS messaging platform as an always-on message exchange that is ready to enable inter-service communication.

In this section, we will explore the following NATS features, which will help us address the challenges posed by the microservice architecture in the context of inter-service communication:

- Provides the decoupling required to operate independently
- Supports plaintext messaging for flexible message format handling
- Supports both synchronous and asynchronous communication models
- Better performance with commodity hardware
- Streaming support for better data guarantees
- Easy to implement client-side code

We will look at each of these features in detail in the upcoming sections.

NATS provides the decoupling required to operate independently

Decoupling is a key concept in the microservice architecture, where each microservice should be capable of running independently while sharing data among them. Most of the message broker solutions, such as Kafka, RabbitMQ, ActiveMQ, and NATS, support a publish-subscribe type of messaging model, which allows the message sender (**publisher**) to interact with the message receiver (**subscriber**) in a fully decoupled manner. All the clients need to know is the address of the server and the name of the publishing entity, such as the topic, queue, or subject. Then, the server will intelligently route the message to the relevant subscribers.

This is the same method that's used by NATS to send messages from the publishers to the subscribers. It uses an entity known as the **subject** as the vehicle that carries the message from the sender to the receiver. This makes development much simpler. Given that the publisher and the subscriber are fully isolated from each other, each party can evolve their applications without impacting the other party, so long as they use the same subject name for communication. Also, new subscribers can receive messages without changing anything on the client side for requirements such as load balancing. The following diagram depicts this concept:

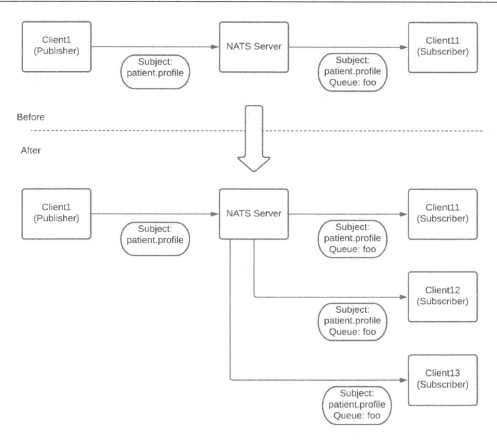

Figure 4.5 – Scaling the subscribers without impacting the publisher

The preceding diagram depicts the initial stage of the solution, where one publisher and one subscriber communicate with the `patient.profile` subject. In this case, the subscriber has subscribed to the subject with an additional parameter for the queue name. NATS provides a built-in load balancing capability with the concept of a queue group. Additional subscribers can subscribe to the same queue name, which follows the same naming convention as the subject. In a scenario where a client starts sending messages at a rate higher than the subscriber can process, it will stack up the messages in the broker and delay the process if there is only one subscriber. At this time, platform architects and developers can add more subscriber clients to share the load among them, without the publisher knowing about it, by subscribing to the same subject with the same queue name.

Another advantage of NATS's subject-based messaging model is that subscribers can receive messages from new clients without changing anything. This can be achieved with the wild card-based subject naming convention. Let's assume we have a scenario where there is a logging service that needs to trace all the messages that are passing through the services. In such a scenario, the logging service can subscribe to the wildcard subject, >, to listen to all the messages that are going through the NATS server. This is depicted in the following diagram:

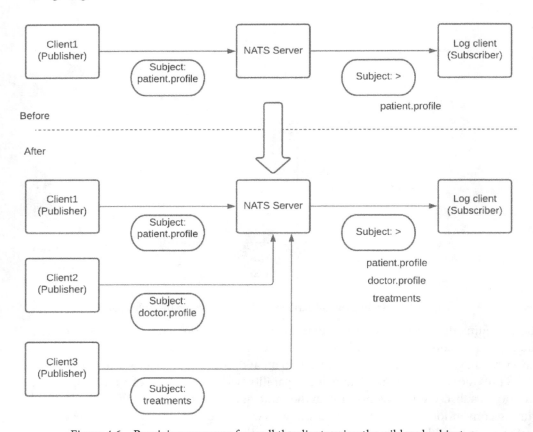

Figure 4.6 – Receiving messages from all the clients using the wildcard subject

The preceding diagram depicts the use case where the initial stage of the platform consists of only one service publishing messages on a subject name called `patient.profile`. These messages are received by the log client (subscriber) with the > wildcard subscription. When more and more clients (publishers) are deployed to the platform, the log client subscriber will automatically receive messages from these new publishers with the subscription.

In a microservice architecture, NATS allows services to be developed with the most suitable technology, and all that service needs to know is when to communicate with other services using the relevant subject names based on the use case. Hence, the applications can be designed using more pragmatic names rather than having to worry about all the redundant text that appears in a REST interface or SOAP interface. This provides more clarity when architecting the solutions and understanding them later. This polyglot type programming model is well supported by NATS via its client libraries and the simple protocol, which can be implemented quite easily by new clients:

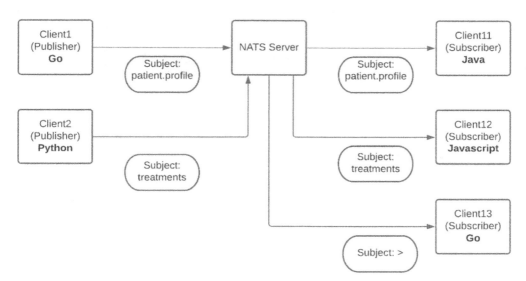

Figure 4.7 – How NATS supports polyglot microservices

The preceding diagram depicts how different parts of the application are implemented as separate microservices using different programming languages. However, all the services communicate with each other using subject names, which do not have any correspondence to the technology they use underneath. A possible example use case would be that JavaScript clients are developed by the user interfaces, which show certain updates on the treatments, while Go clients are developed for logging and tracing. Similarly, the best technology can be utilized for each service with this approach.

NATS supports plaintext messaging for flexibility and simplicity

NATS protocol uses a plaintext messaging format to share data between publishers and subscribers. This allows applications to use a wide range of messaging formats, including JSON, XML, and CSV, when sharing data across services. In a microservice architecture, different services can publish and consume different messaging formats. These services need to process these messages according to the format by using respective algorithms. In typical HTTP-based communication, clients can specify the format of the message with a header such as `Content-Type`. Based on that information, the service that receives the message will do the processing. This is not possible with NATS since it does not share such a header with the message. All it shares is the length of the message in bytes.

This requirement can be fulfilled by using hierarchical subject names in NATS. The following diagram depicts a use case where different services share messages of different formats:

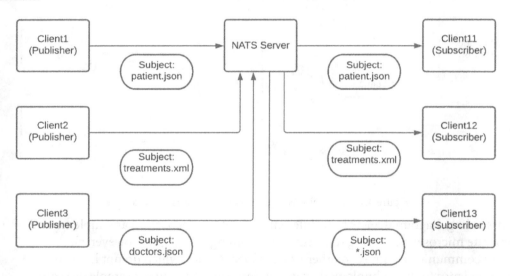

Figure 4.8 – Handling different message formats with subject names

The preceding diagram depicts how to handle different message formats in NATS using subject names that reflect the name of the message format. This provides the publishers and subscribers with awareness of the message formats so that processing these messages is more predictive. By using a single token wildcard, *, certain services can listen to all the messages of a given type, as depicted in the preceding diagram, with Client13, which subscribes to the *.json subject.

Even though NATS offers the flexibility to use different types of message formats, it is recommended to follow one particular format whenever possible for inter-service communication to avoid complexity.

NATS supports both synchronous and asynchronous messaging models

One of the challenges with existing message broker solutions is that designing request-response type systems is not straightforward. Due to this reason, when designing real-time microservices platforms that require synchronous communication, models tend to move away from message broker solutions and go with point-to-point communication methods with technologies such as HTTP or gRPC. NATS comes with support for both the synchronous (**request-response**) messaging model as well as the asynchronous (**event-based**) messaging model so that it can provide the loose coupling required for microservices while supporting real-time system needs:

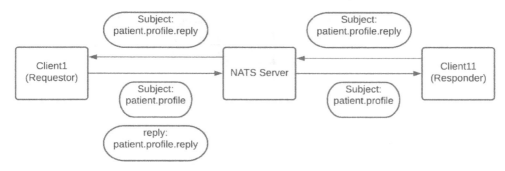

Figure 4.9 – Request-response communication with NATS

The preceding diagram depicts the request-response model of communication with NATS, where the requester sends a message with the subject and a reply subject. This requester will listen for a response on the reply subject and subscribe to it. When the responder receives the message via the subscription to the subject, it will send the response message to the reply subject, which is mentioned in the request. It will publish the message on the reply subject; the requestor is already subscribed to this subject and receives the response. Even though this is a request-response communication, none of the clients (requestor or responder) wait for the other party and operate independently without having any knowledge about each other. That is the advantage NATS brings to the table here.

NATS won't stop there and will allow the requestor to get the best response from multiple responders who subscribe to the subject. This scenario is depicted in the following diagram:

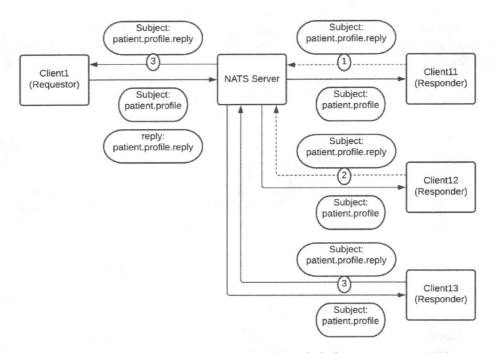

Figure 4.10 – Request-response communication with the best response in NATS

Figure 4.10 is an extension of *Figure 4.9* and here, multiple responders have subscribed to the same subject. All the responders will receive the message, process it, and send the response to the reply subject. The NATS server will receive all three responses and send the first response to the requestor while discarding the latter responses, which are slower. This will help in a use case where a client needs to get a response as quickly as possible from a set of responders.

We discussed the asynchronous communication model of NATS in *Chapter 3, What Is NATS Messaging?*, and you can refer to that chapter for more details on different types of asynchronous messaging use cases supported by NATS.

NATS has better performance with commodity resources

In *Chapter 3, What Is NATS Messaging?*, we discussed the history of NATS and how it evolved as a messaging framework. We also discussed the improved performance it gained by migrating its implementation to the Go programming language, with it ending up processing 18 million messages per second with the latest releases. This high-performance aspect of NATS allowed architects and **Chief Officers (CxOs)** such as the **Chief Executive Officer (CEO)**, **Chief Technology Officer (CTO)**, and **Chief Information Officer (CIO)** to build large-scale software systems with a simple NATS cluster of three servers.

In a microservice architecture, services can scale up automatically when sudden spikes in load are observed with consumer usage patterns. As an example, an IT platform that's been designed for the retail industry can get hit with an unexpected load during periods such as Christmas, Black Friday, and the end of the year. In such situations, having a high-performing messaging platform such as NATS can reduce the burden on operations teams since the NATS server can support these peak loads with minimal changes to the topology.

NATS Streaming provides advanced data guarantees

When NATS was creating, the *at most once* delivery mode was sufficient for the inter-service communication use cases it has been utilized for. Even to date, NATS's core, which is the NATS server that we have been discussing so far, supports this *at most once* guarantee, and most microservice-based systems work well with this model. But in scenarios where advanced message guarantees such as *at least once* are required, NATS Streaming is capable of providing that support. NATS Streaming is useful in scenarios such as the following:

- When old messages need to be replayed.
- A message needs to persist until the subscribers come online.
- Services that consume data at their own pace.
- The data in the message have a **Time To Live** (TTL) beyond the application.

These sorts of requirements are not so common in real-time applications that most of the microservice platforms have been designed for. But there are certain edge cases where both core NATS and NATS Streaming need to be used in conjunction to provide the best of both worlds.

However, NATS Streaming is now deprecated and no further development is happening. Instead, JetStream has been introduced as the successor to NATS Streaming with advanced capabilities around streaming. JetStream is simply the same NATS server running with the additional -js flag to enable JetStream's capabilities. The following features are supported by JetStream:

- At-least-once delivery and exactly once within a window
- Store messages and replay by time or sequence
- Wildcard support
- Encryption for data at rest
- Horizontal scalability

We will learn more about JetStream in *Chapter 10*, *Future of the Microservice Architecture and NATS*.

NATS clients are easy to implement

In typical enterprise environments, people tend to use as few technologies as possible to avoid unnecessary learning curves, as well as to increase the expertise on a selected set of technologies so that those technologies can be utilized well. But that sort of thinking is no longer popular in modern enterprises. Now, people are trying to use the right tool for the right task, irrespective of whether it will be a new technology to the organization, and there is a learning curve for developers. That is why supporting a polyglot programming model in a microservice architecture is important. In such a scenario, where different parts of the application are developed as microservices implemented in different programming languages, NATS will become the messaging backbone that all of these services will interact with for inter-service communication. Having an easy-to-use client interface is critical in this situation since these services must be capable of implementing inter-service communication without much hassle.

In *Chapter 3*, *What Is NATS Messaging?*, we discussed that NATS has a simple protocol with only a handful of command messages to communicate with the server. This simplicity of the protocol makes it easier to implement NATS clients for applications. In addition to that, NATS has a growing community and some people contribute to NATS clients for different programming languages. At the time of writing, there are 40+ clients available for various programming languages, including the following:

- Go
- Java

- Python
- Ruby
- C
- C#
- JavaScript
- TypeScript

These client implementations are available in the following GitHub repository: `https://github.com/nats-io`.

We have provided a set of examples of NATS clients that can be used when implementing microservices with NATS.

The following is a Go client example for connecting to the default server and publishing a message on the `patient.profile` topic:

```go
package main,
import (
    "log"

    "github.com/nats-io/nats.go"
)

func main() {
    nc, err := nats.Connect(nats.DefaultURL)
    if err != nil {
        log.Fatal(err)
    }
    defer nc.Close()

    // Publish a message on "hello" subject
    subj, msg := "patient.profile",
        []byte("{\"name\":\"parakum\"}")
    nc.Publish(subj, msg)
    nc.Flush()
    if err := nc.LastError(); err != nil {
        log.Fatal(err)
```

```
        } else {
                log.Printf("Published [%s] : '%s'\n", subj, msg)
        }


}
```

The preceding code sample can be used with a Go microservice to connect to a local NATS server running on port 4222.

The following is a Java client example for sending a message to a demo server:

```
Connection nc = Nats.connect("nats://demo.nats.io:4222");
nc.publish("patient.treatments", "{"tablets":[panadol,
asithromizin]}".getBytes(StandardCharsets.UTF_8));
// Make sure the message goes through before we close
nc.flush(Duration.ZERO);
nc.close();
```

The preceding code sample connects to the NATS demo server running on the internet and publishes a JSON encoded message on the patient.treatments subject.

The following is a Python client example of receiving a message with a wildcard subscription:

```
import asyncio,
from nats.aio.client import Client as NATS
from nats.aio.errors import ErrConnectionClosed, ErrTimeout,
ErrNoServers

async def run(loop):
nc = NATS()

await nc.connect("nats://localhost:4222", loop=loop)

async def message_handler(msg):
subject = msg.subject
reply = msg.reply
data = msg.data.decode()
print("Received a message on '{subject} {reply}':
```

```
  {data}".format(
subject=subject, reply=reply, data=data))
```

The preceding code imports the required dependencies and makes a connection to the NATS server. After that, it defines a message handler to handle the message once it is received by the subscriber. In this example, we just printed the message data with the subject name and reply subject name.

In the second part of this example, we are creating a subscription on a subject name and publishing a couple of messages to two different subject names:

```
# Simple publisher and async subscriber via coroutine.
sid = await nc.subscribe("patient.*.json",
    cb=message_handler)

# Publish 2 messages.
await nc.publish("patient.profile.json", b'Hello')
await nc.publish("patient.treatments.json", b'World')

# Terminate connection to NATS.
await nc.drain()

if __name__ == '__main__':
  loop = asyncio.get_event_loop()
  loop.run_until_complete(run(loop))
  loop.close()
```

So, you can see how this code example connects to a local NATS server and listens on a wildcard subject to listen to all the messages coming in, with a JSON encoding related to the patient. This example also includes some sample code to publish the messages to the server so that testing can be verified. Next, we look at the advantages of using NATS for inter-service communication.

Advantages of using NATS for inter-service communication

NATS was designed with a purpose. That purpose is inter-service communication. All the design decisions and performance improvements that have been made to the NATS ecosystem were done while keeping this core feature in mind. Before NATS, message broker solutions were mainly used for asynchronous, semi real-time, batch-based use cases and were considered as legacy components. But with the introduction of NATS, these traditional views on the message broker solutions were dismantled and people started utilizing it for their main real-time applications and use cases. The following is a list of advantages that NATS provides that can be utilized in a microservice architecture for inter-service communication:

- NATS supports both request-reply and event-based communication models to support designing applications for both real-time and semi-real-time use cases.

- Support for 40+ client types allows microservices to follow the polyglot programming model without worrying about inter-service communication.

- Multiple levels of message delivery guarantees, including at most once, at least once, and exactly once, provides the quality of services required by most applications.

- Support for multi-tenancy allows NATS to be used in highly scalable, microservices-based applications with decentralized security.

- The possibility to implement security mechanisms such as TLS, key-based authentication, username/password-based authentication, and token-based authentication allows microservices to be implemented with the required security capabilities according to the application.

- The ability to persist the messages via memory, file, or database allows microservices to have improved resistance to failures and data loss.

- Excellent performance and high availability, along with self-healing capabilities, provide the trusted backbone for microservices to communicate at difficult times such as peak usages and node failures.

These advantages highlight the importance of NATS in designing distributed systems, and especially when designing a microservice architecture. We will discuss how these advantages can play a major role in designing practical microservices-based platforms in *Chapter 5*, *Designing a Microservice Architecture with NATS*, and *Chapter 6*, *A Practical Example of Microservices with NATS*.

Summary

In this chapter, first, we discussed the challenges that are faced in microservice-based application design and implementation. Then, we discussed the possible approaches that we can follow to tackle those challenges, especially using the NATS messaging platform. We used several examples to showcase the functionality of NATS using different programming languages. We discussed one of the most important aspects of the microservice architecture, which is inter-service communication, in detail. We went through the specific features of NATS that can help you design and implement successful microservice-based applications and platforms by handling inter-service communication. Finally, we looked at the advantages of NATS, especially in the microservice architecture context.

Now, you should be able to understand the importance of inter-service communication in a microservice architecture and how NATS can be utilized for this. We will discuss a practical reference architecture for microservices by using NATS in the next chapter.

5
Designing a Microservice Architecture with NATS

The microservice architecture is an evolving architectural pattern. Enterprise architects design different solution architectures to solve business use cases with the microservice architecture. Even though those solution architecture patterns are different in many aspects, there are commonalities in these architectures that we can identify. By doing so, we can produce a couple of generic architecture patterns that can be used in most microservice-based platforms. These solutions deal with distinct types of users, systems, and communication models. The communication between these disparate components (humans, systems, applications) is the core of any distributed system, including microservice architecture-based systems.

Inter-service communication is a key aspect of the microservice architecture, and we discussed the different options that are available for architects in the previous chapters. In *Chapter 4*, *How to Use NATS in a Microservice Architecture*, we discussed the advantages of using NATS for inter-service communication in a microservice architecture. In this chapter, we will discuss how to use NATS in a microservice architecture from an architectural perspective and explain how it fits the overall solution based on the microservice architecture.

We will cover the following main topics in this chapter:

- Understanding the key components of a microservice architecture
- Understanding how NATS is positioned in the microservice architecture
- A reference architecture for microservices with NATS

By the end of this chapter, you will understand the role of NATS in the microservice architecture and how it helps developers and architects build effective microservice platforms.

Understanding the key components of a microservice architecture

We discussed the characteristics of the microservice architecture in *Chapter 1*, *Introduction to the Microservice Architecture*, before discussing the challenges of the microservice architecture in *Chapter 4*, *How to Use NATS in a Microservice Architecture*. With that understanding, we are going to dive into the architectural aspects of microservices in this section.

There are two main architectural aspects within the microservice architecture:

- Inner architecture
- Outer architecture

Let's discuss the details of the inner and outer architectures with an architecture diagram. The following diagram depicts these two concepts in a typical enterprise platform:

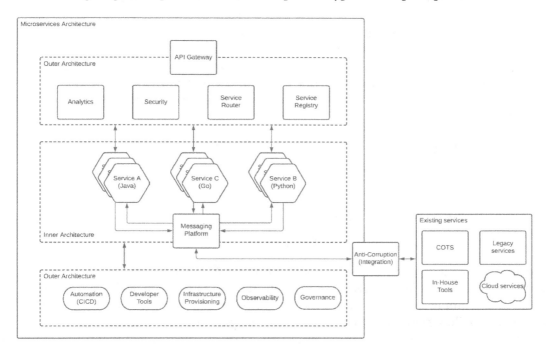

Figure 5.1 – Microservices inner and outer architecture

The microservice architecture is centered around the concept of loosely coupled, domain-driven, independent services. The **inner architecture** covers the implementation aspects of these individual services, while the **outer architecture** covers the wider platform capabilities that are required to deploy, manage, and scale the microservices, along with integration with other systems. We'll discuss these concepts in more detail in the following sections.

Inner architecture

The inner architecture consists of the following components:

- Individual microservices
- Inter-service communication (East-West traffic)

The two components of the inner architecture can be considered as the core of the microservice architecture. The design and implementation of the inner architecture are critical to the scalability and the manageability of the microservice architecture.

Individual microservices

This is the smallest yet most important unit of the microservice architecture. Even though the microservice architecture's design starts with a wider scope or domain (or sometimes with an existing monolithic application), once the scope has been divided into individual microservices, these microservices become independent products or projects. From that point on, it is the responsibility of the (agile) team that has been assigned to work on the microservice to take it to production and maintain it during its lifetime. These microservice teams can decide on the technical aspects of the microservice, such as the following:

- Programming language
- Libraries and/or frameworks
- Design patterns
- Development and testing tools
- Methodology (for example, waterfall or agile)

The first step of building the individual microservices is to select the programming language based on the need at hand and the team at hand. Some team leads would stick to the programming language that was used by the organization in the past and has the expertise to build services quickly. But this option may not be the best since certain programming languages work best for certain use cases. But the risk of this option is that there can be a steep learning curve, as well as the time it takes to deliver the services. It is a matter of weighing the pros and cons of both approaches, starting with an option, and committing to it. If things won't work out, the team can always switch over to the other option since there is no dependency from other teams.

In the world of software development, we do not need to build everything from scratch. There are many great open source and proprietary libraries and frameworks available for building most enterprise applications. Doing proper research and adopting such a library or framework would help teams build applications on top of already matured software components and reduce time spent on building those components from scratch. This will eventually increase the speed of delivery and the quality of the services that are developed.

Design patterns are common knowledge when creating software applications with best practices that are created based on the experience of building similar applications in the past. These patterns help developers get started with the development process with proper guidance that can be followed by different stakeholders of the team.

A key part of development is the tools that the developers use to write and test code. The usage of proper tools such as an **integrated development environment** (**IDE**) helps developers increase productivity and follow a common set of development and security best practices during the development process. By using the test-driven development methodology, developers can build software that can release more frequently with improved quality.

The methodology that's used to release software applications is a core part of any enterprise IT team. Two common approaches are used to deliver applications in the developer community:

- **Waterfall methodology**: In the waterfall approach, applications are developed as a set of sequential steps that are followed one after the other. In this method, the entire application needs to go through the different phases as a whole. As an example, the full microservice needs to be developed first before it's passed to the quality team for testing. Once the testing is complete, the application will be handed over to the operations team so that it can be deployed to the production environment. Once the deployment is complete, the support team will take on the responsibility of supporting the application.

- **Agile methodology**: In the agile methodology, the application is released with small improvements. Each small improvement goes through the aforementioned steps rather than waiting for the entire application to go through a single step. In addition to that, the agile methodology promotes the automation of certain tasks such as build, test, and deployment through automated pipelines, which can be automatically triggered based on code commits.

Based on the use case, the teams can decide on various aspects and work independently. While having this freedom is good for the teams, there needs to be some form of governance across these teams so that they adhere to a common set of values while producing these microservices. This is called **microservice governance** and it is a highly underrated concept in the microservice world. It will be discussed as part of the outer architecture in the upcoming sections.

Inter-service communication

The other critical component of the inner architecture is **inter-service communication**. Even though individual microservices can be designed and implemented as separate products or projects, they should agree when it comes to inter-service communication. This can be one of the following:

- A service definition (schema) in the case of **REST** or **GraphQL**

- A message (protocol buffer) in the case of **gRPC**

- A topic, queue, or a subject in the case of a messaging platform such as **NATS** or **Kafka**

Each of these mechanisms has advantages and disadvantages and we discussed some of them in *Chapter 2, Why Is Messaging Important in the Microservice Architecture?*. In this section, we are not going to reiterate the same details. Instead, we will look at some of the common features of individual microservices, as well as a communication medium that helps with building effective microservice-based platforms.

Failure handling

One of the most important aspects of inter-service communication is handling failures. The microservice architecture is not much different than any distributed system and the common pitfalls of the distributed system exist in the microservice architecture as well. Some common failures are as follows:

- **Network failures**: An entire network or a part of the network can go down due to hardware, software, or power failures. Network failures are common and can happen at any given time. The services need to have proper retrying and circuit-breaking mechanisms to gracefully handle these failures. In most cases, these failures are temporary and are fixed quickly unless they are major failures due to a natural disaster.

- **Server failures**: Sometimes, the computers that are running these services can go down due to hardware, software, or power failures. Server failures can happen when there is a sudden increase in usage or due to bad capacity planning when designing the system and allocating resources. But these kinds of situations can occur, and the system should be capable of withstanding such failures. These situations can cause major outages of the system and with a microservice architecture-based platform, the system can still operate with limited functionality since the functional capabilities are distributed across multiple services.

Concepts such as loose coupling can help in these situations since services can run by themselves without the need for any other services. These failures can last a long time, and the system should operate with limited functionality and avoid cascading failures via mechanisms such as circuit breakers in case of direct communication between services.

- **Service failures**: Individual microservices can go down due to product bugs or performance issues. Service failures are somewhat common, and each service should be able to run with service failures. An individual service should be designed in such a way that it can recover from failures with workarounds such as service restarts, or even with a previous version of the service by rolling back the troublesome service until it is fixed. The other services that communicate with that failed service should be able to function without an outage, sometimes with limited functionality that is missing due to the failed service, but there shouldn't be a total blackout.

- **Scalability failures**: There can be situations where services can fail due to unexpected loads, and this can cascade and cause the entire platform to fail. Scalability failures are not that common unless there are mistakes in the capacity planning or an unexpected usage growth. But we cannot throw away these failures since these are also possible in the real world. The ability to automatically scale the platform is the best possible solution for such scenarios. The microservice architecture helps with auto scalability since every service can scale independently from the others.

The aforementioned types of failures are inevitable, and we must always design the platform while expecting these failures to happen at any given time. But handling these failures can be different based on the application we are building. As an example, in the messaging world, there are different types of delivery guarantees. A few examples are as follows:

- **Exactly once**: Exactly once delivery is used when there is a need to deliver a message to the recipient. This can be useful in an application such as a banking and financial services application. This sort of delivery comes with a cost in terms of the performance of the overall solution since there are more things to take care of to provide this level of guarantee.

- **At least once**: At least once delivery is used in applications where there is no hard requirement to send exactly one message to the recipient(s). In this case, the client (recipient) will take care of the additional messages that are delivered occasionally.

- **At most once**: At most once delivery is the least reliable mode of communication but the simplest to implement and comes with the best performance. In this mode, the messages can be lost during communication, but messages won't be duplicated.

Service discovery

Another important aspect of inter-service communication is service discovery. This is much more critical in a microservice architecture, which is often deployed in container-based deployments. In such an environment, these containers get killed and restarted automatically by the container management platform due to various policies that are enforced by it. We call this uncertain nature of containers *ephemeral*, meaning that it lasts a short time. In such cases, the URL of the services can be changed since each time a container restarts, it will be assigned a new IP address. There are container orchestration platforms such as Kubernetes that can hide this complexity by providing virtual names, called services, to represent these microservices so that the underlying ephemeral nature of the containers is not visible to other microservices.

If a solution consists of a messaging platform such as NATS, this problem does not exist since services are always decoupled from each other via an intermediate component such as a queue, topic, or subject. Each service communicates with the messaging platform, and it is only aware of the existence of that and the abstractions provided by the intermediate components.

Communication mode

Different applications have different needs when it comes to processing data and producing results. As an example, an e-commerce website can have the following requirements:

- Browsing the available items
- Ordering an item

From the aforementioned basic requirements, browsing the available items is the most common use case and depending on the popularity of the platform, there can be millions of people using this functionality. Everyone wants to search for what they need, and the result should be visible in the interface as soon as possible. This sort of requirement can be fulfilled with a synchronous (**request-response**) type of communication model. There can be several microservices sitting in the backend that produce the results for a given interaction by a user. This is an example of a **North-South** message flow, which we discussed in the previous chapter.

On the other hand, ordering an item is a process that takes a while to complete. Because of that, this part of the requirement can be fulfilled with an asynchronous (**event-based**) communication mode, usually with the help of a message broker. In this case, there can be several microservices running in the backend that take care of various activities, such as the following:

- Accepting the order
- Accepting the payment
- Handling the warehouse
- Handling the shipment
- Handling the delivery
- Handling the feedback and review process

These microservices can communicate with each other via an asynchronous mechanism using a messaging platform such as NATS. This is just an example scenario where we need to have both a synchronous mode of communication as well as an asynchronous mode of communication in the sample application. When selecting a messaging platform for inter-service communication, it is good to consider this aspect as well.

Outer architecture

As the word *outer* stands for something outside the core functionality, the outer architecture portion of the microservice architecture discusses the aspects that provide the connectivity to the outside world, as well as the collaboration between microservice teams. As depicted in *Figure 5.1*, there are two main sections of the outer architecture. Those sections are responsible for the following tasks:

- Interacting with the outside world (North-South traffic)
- Governance of the microservice teams and processes

Let's take a closer look at each of these aspects in detail.

Interacting with the outside world

Most of the platforms that are built today interact with external parties via different delivery channels. This allows the businesses and organizations to reach out to a wider audience, as well as grow their business to a global scale. At the same time, organizations cannot live by themselves when doing business globally. Those organizations need to interact with partners, dealers, and other supportive organizations to scale the business to the next level. All this means is that it is an unavoidable requirement to expose the internal business data and processes to the outside world.

In a microservice architecture, we use a component called an **API gateway** to interact with the outside world. It acts as the gatekeeper for the inner architecture where individual microservices and other enterprise systems reside within the enterprise ecosystem. Here are some of the basic functionalities of the API gateway:

- Allow external users to access internal business services (APIs)
- Verify the authenticity (who the user is) of the requests
- Verify the authorization (what they can do) of the requests
- Provide fair usage to everyone who has the right to access the services
- Control excessive usage with rate-limiting and throttling policies
- Monitor the usage of services
- Allow services to be exposed via a standard interface (such as Swagger, Open API Specification, AsyncAPI, or GraphQL)
- The ability to scale based on the capacity requirements

In addition to the aforementioned core features, there are some other aspects of the API gateway that can be considered secondary features. Some of these secondary features are as follows:

- Performance improvements with caching
- Ability to expose legacy services (such as SOAP)
- Ability to expose asynchronous services (such as JMS, Kafka, and NATS)
- Flexible deployment model (on-premises, cloud, and hybrid models)

When interacting with external consumers, there are additional components that can help the API gateway build a better experience. As depicted in *Figure 5.1*, the following components can help an API gateway build a great experience for the outside consumers of the microservice architecture:

- **Service Registry**: This is the component that is used to allow consumers to find the services (APIs) that they can use to consume the services offered by the enterprise. These users are not the end users who consume the business services via a mobile or web application. These are the developers who use these services to build mobile or web applications that end users consume. These developers can be internal or external.

  The service registry comes with a set of capabilities that increases the productivity of the developers and the overall quality of the services offered by the enterprise. A few such capabilities include browsing services, searching for services, subscribing for consumption, trying out examples, providing feedback to API owners, community interactions (discussions, rating, and sharing), and much more. In the context of API management, this component is sometimes called a **Developer Portal** since it is used mainly by the developers.

- **Service Router**: This is the functionality that allows the API gateway to execute certain business logic such as service orchestration, message transformation, connectors to integrate with cloud and on-premises systems, and routing based on headers and content.

  Depending on the state of the enterprise system at hand, this service router can execute certain intermediate business logic execution before exposing it to the consumers via the API gateway. This component can be a fully-fledged **enterprise service bus** (**ESB**) or a lightweight mediation engine that can only perform a selected set of operations, such as the ones we mentioned previously. Even though we depicted this as a separate component inside the outer architecture, some API gateways contain this functionality within the gateway itself.

- **Analytics**: Business leaders and CXOs always keep an eye on the **Return on Investment (ROI)** side of the IT projects. Hence, having proper **Key Performance Indicators (KPIs)** and measuring them against the API program is crucial to the success of any API program, and **analytics** plays a major role in this regard by providing the required business intelligence and statistics on the behavior of the APIs and its usage. Some of the details that are captured by the analytics component are as follows:

a) How is each API used overall (requests per second, minute, or month)?

b) What is the latency of each API?

c) Who are the consumers of the APIs?

d) What are the popular APIs?

e) Which APIs are failing frequently?

There can be many other statistics and metrics available through the analytics component in a real analytics tool. The purpose of this component is to identify the behavior of the services and make key business decisions based on that.

- **Security**: This is one of the critical capabilities offered by the API gateway when exposing services to external consumers. Since security is a broader domain of its own, having a separate component to implement security helps API gateway offload the complexity that comes with securing services. Some of the challenges that are faced when securing services include having a lack of a standard protocol when securing services, needing to integrate with many different user stores and identity providers, and needing to provide advanced security capabilities such as **two-factor authentication (2FA)**, social login, and passwordless authentication.

Due to these and many other reasons, having a separate component to work with the complexities of security requirements makes things scalable and manageable for the API gateway. There are specialized products that can provide this capability, and these products are commonly known as **identity providers (IDPs)**.

Governance of the microservice teams and processes

In *Figure 5.1*, there is a bottom section that also relates to the outer architecture. It covers the various operational and governance aspects of the microservice architecture. There are several components mentioned in this diagram that we will take a deeper look at in the coming sections. These are as follows:

- Automation
- Developer tools
- Infrastructure provisioning
- Observability
- Governance

Let's discuss these components in detail.

Automation

In the *Inner architecture* section, we discussed that each microservice can be developed by different teams with their own technology choices, such as programming languages, libraries, and design patterns. But when it comes to delivering these microservices as products to the customers, these microservices need to be deployed into the infrastructure that is used by the enterprise. Hence, these teams should adhere to the common standards and technologies that are adopted by the wider enterprise when it comes to deploying these services into a production environment. This automation allows these microservice teams to build software delivery pipelines while utilizing enterprise-wide tools to release microservice frequently, without the need to spend too much time on manual testing and manually provisioning resources in the platform. Sometimes, this automation process is called **continuous integration** (**CI**) and **continuous deployment/delivery** (**CD**), which covers the development, testing, and deployment aspects of the microservices.

Developer tools

Sometimes, we can see that developer tools are related to the technological choices that are made by each microservice team, which is why it is part of the inner architecture. But most of the tools we consider as developer tools can be shared across teams that use different programming languages. A few examples of such common developer tools are as follows:

- IDEs (Visual Studio Code, IntelliJ, and Eclipse)
- Source code management tools (GitHub and GitLab)
- Container management tools (Docker and Kubernetes)
- Testing tools (Postman and JMeter)

These tools can be managed and maintained across microservice teams, so we consider them as part of the outer architecture.

Infrastructure provisioning

In a typical enterprise platform, the underlying infrastructure that's used to run applications is decided at a corporate level due to the size of the investment that goes into it. As an example, the choice of infrastructure impacts the platform for the next 5-10 years. Hence, it is always a corporate decision and the individual microservice teams must work with whatever infrastructure is chosen. Having said that, certain large corporations go with a multi-infrastructure choice due to reasons such as getting competitive prices, avoiding vendor locking, and getting the best of both worlds.

In the technology world, this is sometimes called the **multi-cloud strategy** since most of the infrastructure that's used by organizations appears to be running on the cloud. Even in such an organization, most of the time, a given application will only utilize one infrastructure (cloud) platform. Hence, most of the time, infrastructure provisioning is part of the outer architecture, with central governance across microservice teams.

Observability

The term *observability* is sometimes mixed up with the term *monitoring*. Even though both talk about achieving a similar goal, the concepts are not the same. In monitoring, we actively record information about our systems, and it is an action that we do on top of the systems we run and manage. We monitor applications to detect any failures and anomalies so that we can take the necessary actions to fix those behaviors. But observability is not an action. It is a property of the system. It is a measure of how well the internal states of the system can be inferred from the knowledge of its external outputs. This means it is a measure of how well the applications are designed with external outputs that can be used to derive the state of the application in any situation.

If a system does not have good observability, then even the best monitoring tool will not be able to identify the issues and anomalies of the application. Hence, in a complex microservice architecture with hundreds or thousands of microservices, the quality of the observability is a key factor when identifying failures and recovering from them. Hence, it is essential to use a common observability approach across microservices so that troubleshooting the overall solution is possible. That is why observability is considered a component in the outer architecture of the microservice architecture.

Governance

Microservice governance is a topic of its own. In the previous sections, we discussed how a common approach to automation, developer tools, infrastructure provisioning, and observability can help us build a better microservice architecture. Governance is the process that ties all those aspects together and manages those aspects across microservices. The following diagram depicts the concept of governance in a microservice architecture:

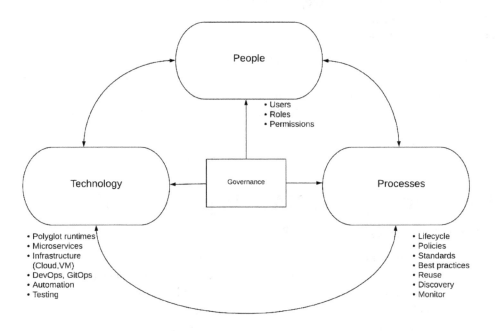

Figure 5.2 – Microservice governance with people, processes, and technology

The preceding diagram depicts the three distinct aspects of microservice governance and what each aspect deals with in the real world. They are described as follows:

- **People**: The microservice architecture suggests making some changes to the organizational hierarchy, which would require the human resources to be managed properly with defined scopes and responsibilities. Microservice governance makes this more systematic by defining the groupings (roles) and respective permissions (responsibilities) that are assigned to the users (people). As an example, a team can be created with a set of individuals with different skills, including development, testing, user experience, and architecture, and this team can be assigned a specific role so that they can make their own decisions to deliver the microservice.

- **Processes**: Even though the microservice architecture allows individual teams to follow their own mechanisms for certain tasks, there should always be standardized processes for certain tasks. These are defined as policies or standards to reflect the process that needs to be followed by the teams. Some policies reflect on technology such as infrastructure choice standards, while some policies reflect on the overall interactions of team members, such as life cycle management and the approval process.

- **Technology**: This component is the delivery vehicle of the services that people develop using the processes we define. All the aspects of governance we implement using people and processes should not hinder the technology and its use. Instead, it should strengthen the technology to offer the best quality service to the consumers.

With this, we have concluded the topic of the inner and outer architectures. Now, we can learn where NATS is placed in the microservice architecture.

Understanding how NATS is positioned in the microservice architecture

In the preceding section, where we discussed the inner architecture of the microservice architecture, we discussed how to use a messaging platform for inter-service communication. This concept is depicted in *Figure 5.1* as well. In this section, we are going to discuss how NATS fits into the role of this messaging platform and how it is going to help the microservice architecture with its capabilities. The following diagram depicts how NATS can be used in a microservice architecture as a messaging platform:

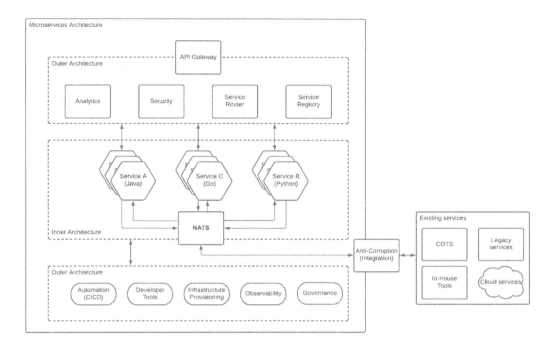

Figure 5.3 – The microservice architecture with NATS

The preceding diagram depicts how NATS fits into the inner architecture of the microservice architecture and how it interacts with the other components. The following are the major interactions between the NATS server and the remainder of the microservice architecture:

- Microservices interact with the NATS server as NATS clients.

- Existing services interact with the NATS server via the anti-corruption layer.

- Security components interact with the NATS server when securing the service interactions using advanced security mechanisms such as JWT.

Let's discuss the interactions between the NATS server and the other parts of the microservice architecture in detail.

Microservices interact with the NATS server as NATS clients

Microservices need to interact with each other and share data between them at certain times. We identified this interaction as inter-service communication, and the NATS server is used as the intermediate component that is used for this purpose. It helps to decouple the communication between microservices. Each microservice acts as a client to the NATS server, regardless of whether it is the message sender (**publisher**) or the receiver (**subscriber**). All the microservices are considered the same by the NATS server, and any microservice can publish messages or subscribe to messages with the client APIs provided by the respective client libraries. The following diagram provides a detailed view of the interaction between the microservices and the NATS server in a container-based Kubernetes deployment:

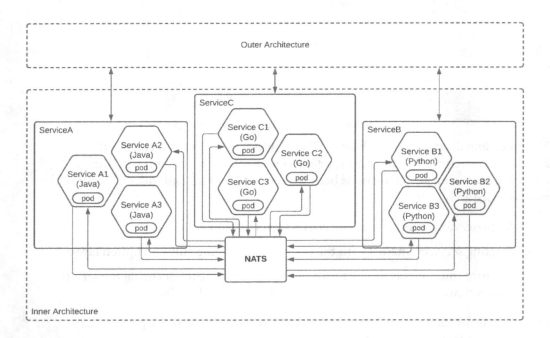

Figure 5.4 – Microservice inner architecture with the NATS server

The preceding diagram depicts a scenario where microservices have been deployed in a Kubernetes-based infrastructure. It consists of three microservices that have been implemented in the Java, Go, and Python programming languages. Each microservice has three instances (replicas) that have been deployed for scalability, and each instance is deployed as a pod in the Kubernetes environment. The pods that are related to the same microservice are exposed as a service to the outer architecture layer. Since this is the first time we are mentioning Kubernetes, let's explain the Kubernetes concepts that are being used here:

- Kubernetes is a container orchestration system that takes care of deploying containers across multiple physical or virtual computers. It provides a singular interface to the application developers to deploy their applications as containers, without worrying about the underlying infrastructure's complexity.

- The pod is the smallest unit of deployment in Kubernetes and it can run one or more containers that are related. Pods also have IP addresses. In this example, each microservice has been deployed as a container within a pod.

- Service is a method that's used in Kubernetes to expose a set of related pods as a single interface to the consumers. It load balances requests across the pods that are running underneath.

If you are completely new to the concept of containers and Kubernetes, you can refer to the official Kubernetes website (`https://kubernetes.io/`) to learn more about it.

Even though each microservice is exposed to the outside world through services, internally, there are multiple replicas of the microservices running. As an example, Microservice A, which is implemented in Java, has three instances running as separate pods, and each of those microservices interacts with the NATS server via the Java client interface. Since these are identical replicas of the same microservice, each service can be published to the NATS server with a given subject name and receive messages via a subscription with a given subject name.

Each microservice has a communication channel for publishing and subscribing, as depicted in *Figure 5.4*. The same concepts are applied to the other microservices; that is, Microservice B and Microservice C. Microservice B has three instances written in Python, and each of those microservices instances communicates with the NATS server using the Python client API. This concept can be extended to hundreds of microservices with a different number of replicas, and the NATS server can handle the communication between microservices without creating a complex spaghetti architecture inside the enterprise platform.

Existing business applications and services interact with the NATS server via the anti-corruption layer

Even though we prefer to build an entire enterprise platform with a microservice architecture, many systems were built before the microservice project, which cannot be replaced or rearchitected due to their importance to the overall business. These types of enterprise platforms, which have both microservice-based systems and non-microservice-based systems, are referred to as **brown-field** enterprise platforms.

If a particular enterprise platform only consists of microservices, such platforms are called **green-field** enterprise platforms. When two distinct types of systems exist in an enterprise platform, the interaction between those systems can be done through an intermediate component. This is often referred to as the **anti-corruption layer** pattern in the software architecture. There are tools such as **enterprise service bus** (**ESB**) and the integration platform that can translate the communication between two systems.

The following diagram depicts how the interaction can be done between the microservice architecture and the existing enterprise applications:

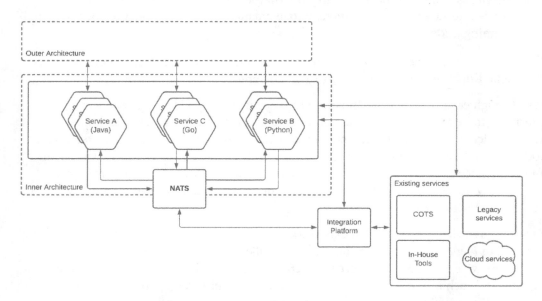

Figure 5.5 – The microservice inner architecture interaction with existing enterprise systems

As depicted in the preceding diagram, there are existing business applications and systems that were used by the organization to fulfill the desired business and IT needs. A few example systems are depicted in the preceding diagram. These are as follows:

- **Commercial Off The Shelf (COTS)**: COTS systems are turnkey business applications that provide certain business functionality. A few examples are **Customer Resource Management (CRM)** and **Enterprise Resource Planning (ERP)** systems.

- **Legacy services**: Certain legacy applications and services can be purchased during the early days of the business that haven't been replaced or upgraded.

- **In-house tools**: Many applications and services can be developed by internal IT teams to execute certain day-to-day business operations.

- **Cloud services**: It is hard to find an enterprise organization that does not use cloud services that offers **Software as a Service (SaaS)**.

These internal systems must communicate with the microservice architecture. But these systems may not speak the same language as microservices. In other words, these systems may use different wire-level protocols and messaging formats that are proprietary in some cases. Due to this disparity, an intermediate component such as an integration platform that can transform message formats and translate wire-level protocols can be used as the anti-corruption layer. These microservices can communicate with these external systems via one of three channels:

- Communicate over NATS server via an integration platform

- Communicate directly via an integration platform

- Communicate directly with the existing systems

Since microservices are communicating with each other using the NATS server, whenever there is a need to communicate with an existing business application, it can use the NATS server for that purpose as well. If there is a need to decouple the interaction between the microservice and the enterprise system, this method is suitable. In this case, the NATS server will communicate with the enterprise application via the integration platform, which will translate the NATS message into the relevant protocol.

If the microservice needs to communicate with these business applications synchronously, it can communicate over the integration platform if there is a need for protocol or message format transformation. If not, the microservice can directly communicate with the business application without going through the integration platform. Both of these options are depicted in *Figure 5.5*.

Security components interact with the NATS server for implementing advanced security

NATS allows you to secure communication with the server using multiple options. We discussed these options in *Chapter 2, Why Is Messaging Important in the Microservice Architecture?*. It provides several options for authenticating the clients when connecting with the server. These options are as follows:

- Tokens
- Username/password (basic authentication)
- TLS authentication (certificate-based)
- **NATS key** (**NKey**) with a challenge
- Decentralized JWT-based authentication/authorization

All these options except JWT-based authentication work without any interaction with an external security component. All these options can be configured in the NATS server itself; the clients need to provide the required authentication details when connecting with the server based on the authentication type. As an example, with the token-based model, the client connects to the server with the token attached to the connection request. Then, the server validates the token with the information that has been configured within the server.

With the decentralized JWT-based authentication model, the NATS server requires an external security component to manage the users and accounts. Before discussing JWT-based security, let's briefly understand what a JWT is.

JSON Web Token (JWT)

JWT is a mechanism that is used to represent claims to share between two parties. It is an open standard defined with **RFC 7519** (`https://datatracker.ietf.org/doc/html/rfc7519`). **Claims** are used to identify information on a subject. Standard JWT claims are digitally signed and verified. NATS uses a trust chain to validate the users with JWT. In that trust chain, there are three main components:

- Users
- Accounts
- Operators

These are configured as roles in the JWT claims. These roles form a hierarchy. Operators issue accounts, while accounts issue users. If an account is issued by an operator that is trusted by the NATS server, account users are trusted.

When a user (client) connects to a server using the JWT-based authentication mode, it presents a JWT issued by a given account. Then, the server will issue a cryptographic challenge to the user that the user will sign with its private key. Then, the server validates the signature using the user's public key. If that is successful, then the server retrieves the account details that are mentioned in the JWT and verifies the user issuer with that account. Finally, the server checks whether a trusted operator that's been configured with the server has issued the account. If that is successful, the connection is allowed.

The JWTs for users, accounts, and operators are managed separately with the tools provided by the NATS ecosystem. This allows us to manage the JWTs in a separate location that is independent of the NATS server. These are saved as files and folders. With the automation tools that are part of the outer architecture, these tokens can be managed in an automated manner, without the need to change the server configurations when you must create/update or delete these tokens and security configurations. We will discuss securing NATS server interactions with JWT in more detail *Chapter 7, Securing a Microservices Architecture with NATS*.

In addition to these three main interactions, the NATS server interacts with the outer architecture components, such as infrastructure provisioning and observability. Infrastructure provisioning components interact with the NATS server while the server is being deployed and maintained. The observability component interacts with the NATS server when the NATS server is being monitored.

With that, we have discussed the microservice architecture components and how NATS can interact with these components in a real-world deployment. Now, let's put all this knowledge together and produce a reference architecture that can be reused to build an effective, scalable, and manageable microservice platform.

Reference architecture for microservices with NATS

The microservice architecture is an evolving architecture pattern. This means that with every iteration, you will find that something has changed in the architecture. Some components may have been added, some components may have been modified, and some components may have been removed. Yet, the fundamental concepts stay the same. Due to this, we can come up with a common reference architecture that captures the core concepts of the microservice architecture.

In this section, we will produce a reference architecture for microservices while using NATS as the inter-service messaging backbone. This architecture will capture the core components of the microservice architecture, along with the integrated components that you find in enterprise distributed systems. The following diagram depicts a comprehensive microservice architecture with NATS:

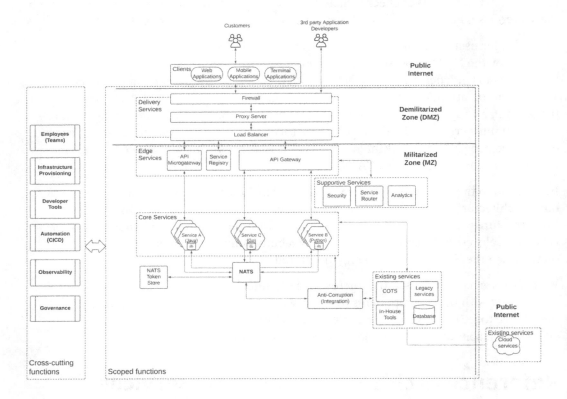

Figure 5.6 – Microservice reference architecture with NATS

The preceding diagram captures a lot of details that we have not discussed yet, as well as some details we discussed in the previous sections of this chapter. Let's discuss the main components that are depicted in the preceding diagram.

Clients

The end goal of building enterprise information technology systems is to provide better experiences to the customers. These customers use different clients to access the digital services that are offered through the platform. We can categorize these clients into three main types:

- **Web applications**: These are the clients that access the platform via a web browser. Customers may use devices such as mobile phones, laptops, and tablets to access these services via a web browser. In these web applications, server-side code is stored in web servers that are running on the same enterprise platform in front of the API gateways.

- **Mobile applications**: These are the native mobile applications that access the platform directly via API calls. These clients do not use browsers so no web server is required on the enterprise platform to host the server-side of the web application.

- **Terminal applications**: These are headless applications running on computers that directly access the system via APIs. These applications are similar to mobile apps and do not use a browser to access the services.

Delivery services

Enterprise platforms offer valuable digital services to the customers via the clients we mentioned in the preceding section. These client applications are available for anyone to use via a URL (web application) or a mobile application that can be downloaded via an application store. Hence, it is essential to control the access to the platform while providing improved services to the actual customers. These delivery services are placed in front of the API gateway component to allow clients to access the internal services via the public internet. These delivery services are placed in a publicly accessible **demilitarized zone (DMZ)**. Three main types of delivery services are used in enterprise platforms:

- **Firewall**: This is a network security system that controls and monitors incoming and outgoing traffic based on a set of predefined rules.

- **Proxy server**: A proxy server is an intermediate component used in networking between the client and the server to hide the internal details of the server that are exposed via the service. In an enterprise, we use a reverse proxy to hide the internal server URL and use the proxy server URL instead.

- **Load balancer**: This component is used to distribute traffic across multiple instances of the same service. As an example, if there are multiple instances of the API gateway that have been deployed for high availability, there should be a load balancer to distribute the load across those nodes.

Not all enterprise platforms have all three components deployed since there are alternative ways to achieve some functionality via the internal components, such as API gateways. Also, some products combine these functionalities and provide a single component. In addition to exposing the services to public clients, these delivery services provide additional features, such as the following:

- **SSL/TLS termination**: Incoming traffic from the clients comes as SSL/TLS protected requests. Terminating transport layer security at this layer improves performance within the internal components.

- **Caching**: Caching at this layer improves the overall experience of the customers by improving response times.

- **Security**: This layer helps protect the business services from various attacks, such as **Denial of Services (DOS)** and **Distributed DOS (DDOS)** attacks.

Edge services

Once the requests have been filtered and load balanced through the delivery services, these requests reach the edge services layer, which is where the API gateways are. These API gateways provide the functionalities that are required at the edge of the platform. There are three types of API gateways you can deploy based on the requirements of the platform:

- **API gateway**: This is the standard mode of deployment for the API gateway, which exposes multiple microservices as APIs. We call this approach the **monolithic gateway**.

- **API microgateway**: In this mode of deployment, only one or a few selected APIs are deployed in this gateway. In this mode, APIs are deployed across multiple gateways in a partitioned manner. Each microgateway can scale independently based on the capacity needs of the clients.

These gateways utilize supportive services to provide the required functionality at this layer. In addition to the API gateways, there is the service registry at the edge services layer that can be accessed by third-party application developers. These developers will utilize the APIs that are exposed by the platform to build new applications so that businesses can scale to new heights. This service registry integrates with the API gateway as the runtime component.

Supportive services

The requests that reach the gateway need to be authenticated, authorized, and throttled based on the defined policies. This functionality is provided through the security component in supportive services. Some enterprise platforms have a dedicated **Identity and Access Management (IAM)** solution that can cater to this requirement. Most of the API gateway vendors provide this functionality as a built-in feature that's supported by integrating with an external IAM solution.

Another key requirement at the gateway layer is monitoring API traffic and making real-time decisions based on traffic patterns. This functionality is provided through a separate monitoring and analytics solution that is provided by the API platform. This component helps business leaders and architects make decisions regarding the API's development and the evolution of existing APIs.

Sometimes, requests that come into the gateway do not adhere to the formats that are expected by the backend microservice layer. In such situations, the service router acts as a mediation layer and transforms the message into the desired format. In addition to that, certain simple business logic such as content-based routing and message tracing can be implemented with this component.

Core services

Then comes the business logic and data layer, where we have the microservices implemented as independent, loosely coupled, domain-driven services. These services provide the core functionality requested by the customers and manage business data. We discussed the implementation of these services in the *Inner architecture* section. These services communicate with each other using the NATS messaging platform. In addition to that, some of these services will communicate with the existing business applications and services through an integration platform or directly. Depending on the communication mode (synchronous or asynchronous), the messaging formats, and the wire-level protocol, this communication will happen directly or via the integration platform.

NATS server

The NATS server provides the always-on, highly performant messaging backbone for inter-service communication. The core microservices communicate via this messaging backbone and integrate with existing applications and systems through the integration platform whenever possible. This makes the microservices and existing services operate independently yet integrate when required. Since NATS provides both synchronous and asynchronous communication models, with the usage of the integration platform for protocol and message translation, microservices can integrate with these services as and when required. The NATS server will store the JWT-based tokens and configurations in a separate file store that sits outside the NATS server. In addition, to act as the messaging backbone for inter-service communication, the NATS server can also process any data that's generated as events from other systems, and it can also integrate those events with the necessary target systems.

Integration platform

Microservice-based implementations need to live with the enterprise applications that were introduced a long time ago. These applications may have different wire-level protocols and messaging formats than the microservices. The integration platform helps microservices communicate with these existing services, as we discussed in the previous section.

Existing services

These are the enterprise applications and tools that are used by the IT teams before the microservices are introduced. These systems and applications play a key role in business and IT operations. These systems hold key business information, and microservices must interact with these systems to build innovative services for the customers in an agile manner. We discussed a few examples of these systems in the previous section.

Cross-cutting services

The different types of services we've discussed in this section require a common set of functionalities. These common functionalities include the following:

- **Infrastructure provisioning**: Every component within the services we discussed requires computing resources to execute their tasks. Hence, it is essential to have a common strategy on the infrastructure choice for the platform. As an example, the project leaders should agree on what components are deployed on what infrastructure based on the needs of that component.

- **Developer tools**: Another capability that is shared by microservice teams in the various development-related tools are security scanning tools such as Veracode and Qualys, testing tools such as Postman and Jmeter, and source code management tools such as GitHub and GitLab.

- **Automation (CI/CD)**: Automation allows the deployments to be more frequent and less prone to human errors. The overall platform should decide on the automation tools such as build servers, test tools, configuration management tools, and infrastructure.

- **Observability**: This is a key aspect of the overall system design since troubleshooting errors in a highly distributed system, such as the one we discussed here, can be impossible without proper observability at each services layer. A common way of implementing observability is via instrumentation at each layer by using a common tool or open standard such as open tracing. In addition to that, having proper logs printed to the log files allows the platform designers to use log monitoring tools to build correlations between log entries using a common ID, such as correlationID, which will be passed through each services layer as a header or an entry to the message.

- **Governance**: The governance process manages the life cycle of development processes and automates certain business operations that would help the individual teams execute their tasks efficiently, without having to wait on human approvals, which would cost them time otherwise.

- **Employees (Teams)**: On top of everything, we need people to design, build, and manage these systems and applications. Hence, it is essential to have a culture that is suitable for developing services in an agile manner that would help to reap the true benefits of the microservice architecture. Instead of having **Center of Excellence (CoE)** teams working in silos, the microservice approach demands teams that are agile and consist of people from different CoE teams to work together in a timebound manner to produce microservices. This needs to come from the top levels and change should occur in the individuals who work on these projects.

That concludes the reference architecture for microservices with NATS. In this section, we covered the high-level architecture and the relevant components in detail. Now, let's summarize what we've learned in this chapter.

Summary

In this chapter, we discussed the key components of the microservice architecture by discussing the inner and outer architecture of microservices in detail. There, we discussed how individual microservice development and inter-service communication becomes the inner architecture. Then, we discussed the outer architecture components at the edge of the platform and the cross-cutting functions that are shared between microservices such as observability, infrastructure provisioning, and automation.

Then, we discussed how NATS can be used in the microservice architecture and went into the details of how the NATS server interacts with the various components of the architecture. Finally, we defined the reference architecture for microservices with NATS by aggregating the points we discussed in the previous sections and discussed each component within the architecture in brief. This chapter helped you understand the core concepts of the microservice architecture and how NATS fits into that architecture when designing an effective solution with microservices. This knowledge helps you make the right choices when we implement an example application with microservices.

In the next chapter, we will use this reference architecture and build a reference implementation of a microservice-based solution for a healthcare application.

Further reading

For more information regarding what was covered in this chapter, take a look at the following resources:

- The Kubernetes documentation is available at `https://kubernetes.io/`.
- The JWT specification is available at `https://datatracker.ietf.org/doc/html/rfc7519`.

6

A Practical Example of Microservices with NATS

At this point, we have a strong understanding of the microservice architecture and the NATS messaging technology. In this chapter, we are going to use our knowledge to build a real-world application using a microservice architecture with NATS as the inter-service communication mechanism. We will use the reference architecture of microservices that we discussed in the previous chapter to implement an application for a healthcare service provider such as a hospital.

The following topics will be covered in this chapter:

- Understanding the use case of a hospital **outpatient department** (OPD)
- Defining the solution architecture
- Implementing microservices
- Setting up the NATS servers cluster
- Trying out the sample application

By completing this chapter, you will have learned how to implement microservices that communicate with each other using the NATS protocol. In addition to that, you will also learn how to set up NATS servers and configure them for a microservice-based application.

Technical requirements

In this chapter, we will be implementing a set of microservices using the Go programming language and set up NATS servers, along with a few other tools. You need to install the following software components to try out the examples mentioned in this chapter:

- The Go programming language
- The NATS server
- JDK 11
- MySQL server

The source code for the examples used in this chapter can be found at `https://github.com/PacktPublishing/Designing-Microservices-Platforms-with-NATS/tree/main/chapter6`.

Understanding the use case of a hospital OPD

In a typical healthcare service provider institution such as a hospital, there is a dedicated department to serve the patients that need treatment for short-term illnesses that do not require a bed or need to be admitted to the hospital immediately. This department is called an OPD. An OPD is an important part of the hospital's operations and provides a buffer between the incoming patients and the wards that treat patients for longer periods. We can identify a certain number of main tasks that are executed within the OPD. We discussed these tasks in *Chapter 1, Introduction to the Microservice Architecture*. These tasks include the following:

- **Patient registration**: This is where new patients are registered with the hospital so that they can be provided with the necessary medical treatment.
- **Patient inspection**: The immediate next step after registering is the patient being inspected by experienced physicians to identify the condition of the patient based on their symptoms and behaviors.

- **Temporary treatment and testing**: Physicians will decide whether the patient needs to be immediately transferred to a ward by admitting them to the hospital. Otherwise, they will give a temporary treatment and/or execute a few additional tests to identify the root cause of the illness.

- **Releasing the patient**: Once the temporary treatment and the test results have been received, the physicians will evaluate the condition of the patient and decide whether to send the patient back home or admit them to the hospital ward for further treatment.

The execution of these tasks involves different types of users, and each user may have a different role to play in the overall functioning of the OPD. Let's identify these users as well:

- **Physicians or doctors**: These are the main stakeholders of the unit who inspect the patients, suggest treatments and tests, and make decisions on releasing the patient.

- **Nurses**: They are involved in providing temporary treatments, medication, and obtaining samples for testing and reporting the conditions of the patient to the physician so that they can make better decisions.

- **Medical staff (other than doctors and nurses)**: They execute certain non-critical tasks such as sample transportation, medication transportation, and helping patients.

- **Non-medical staff**: These are the stakeholders of the unit who keep the unit clean and organized.

- **Patients**: These are the main consumers of this unit who received treatment from the unit.

- **Relatives/guardians of the patient**: These are the supportive people who help the patient.

Now that we understand the tasks that are executed within the OPD and the major actors/stakeholders, let's try to identify the interactions of the actors with tasks and the interactions among these tasks in detail.

Patient registration process

This process allows patients to be registered on the hospital database. If the patient is a first-time visitor to the hospital, then the patient will be registered as a new user. Either the patient or the guardian can provide the details of the patient so that the medical staff can store the information in the database or the **Electronic Medical Records (EMR)** system. In addition to this, if the patient is already registered in the system and needs to modify a certain data field, such as the address or the contact details, that is also possible. In either case, the patient will be issued a token with a number so that the patients are treated on a first come, first serve basis. The following diagram depicts the high-level details of the patient registration service that we will develop in this chapter:

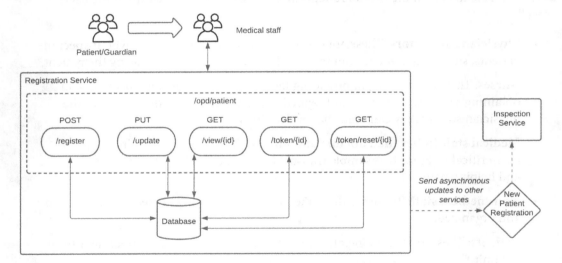

Figure 6.1 – Patient registration service

The preceding diagram depicts the interface that this service exposes to the consumers using a REST-based approach. There is a REST API that is defined with a base context path of /opd/patient. Under this API, there are five resources to execute different functions on the service:

- A resource with the /register context path with the POST method to record all the registrations of new patients.

- A resource with the `/update` context path with the PUT method to update an existing patient with new information. In this resource, the ID field is used to identify the patient uniquely. In a real-world scenario, this can be the **National Identity Card (NIC)** or **Social Security Number (SSN)** of the person.

- A resource with the `/view/{id}` context path with the GET method to view the patient details in case the patient already exists in the system. If the patient already exists in the system, it won't be necessary to register the patient again unless their details have changed, which we can execute through the update resource.

- A resource with the `/token/{id}` context path with the GET method to generate a token for a particular visit for a particular patient. The patient is identified by the `{id}` path parameter.

- A resource with the `/token/reset/{id}` context path with the GET method to reset the token generation after a certain time (daily). If there is an intermediate failure in the token generation, it can be reset to the previous token number by providing the `{id}` path parameter.

The service that is exposed over REST uses a database that is local to the service to store the information and fetch the necessary details.

For interservice communication, this service uses an asynchronous communication model. It uses an event-driven approach to send new registration events, patient data update events, and token generation events so that other services such as inspection services are aware of the new patients coming into the OPD.

From the consumer's perspective of this service, medical staff will speak to the patient or their guardian and capture the details about the patient. Then, they will update the information in the system via the provided web or mobile application interface.

Patient inspection process

Each patient registration sends a notification to the patient inspection process with the patient's details and their token number. This allows the patient inspection process to be synchronized with the registration process. The physicians will call out the number via an automated system using a display and the patient will go to the physician according to the token number. Once the inspection is done, the physician will update the patient inspection report and it will be automatically sent to the temporary treatment unit with the associated patient ID. The following diagram depicts the high-level implementation details of the inspection service:

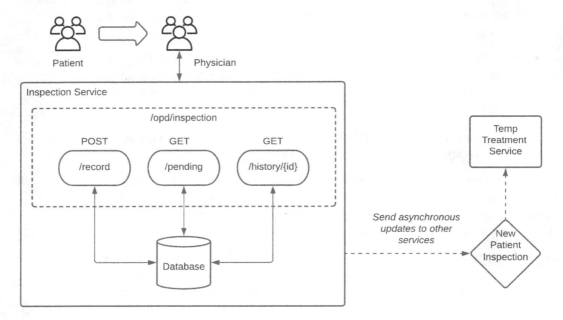

Figure 6.2 – Patient inspection service

As depicted in the preceding diagram, the inspection service exposes an API to interact with external consumers. A REST API is defined under the /opd/inspection context path to isolate this service from other services.

Under this API, three resources are implemented:

- A resource with the /record context path with the POST method is implemented to record the inspection details and the necessary treatments under the patient ID.

- A resource under the /history/{id} context path with the GET method is implemented to view the historical treatments that were given for this patient.

- A resource with the /pending context path with the GET method is implemented to view the pending inspections of the patients that were registered via the registration service.

When communicating with external services, it uses the asynchronous communication method. Patient inspection data and observations/medications are recorded in the local database and sent as asynchronous events to the treatment service.

In this scenario, the external consumers are the physicians who speak with the patient and use this service to record the details. This ensures that the temporary treatment and testing can be done based on the inspection results.

Temporary treatment and testing

Once the inspection is done and the treatments and tests have been recommended by the physicians, the nursing staff will take the necessary steps to treat the patient with the relevant medication and get samples for testing. These treatments are marked against the user with information about the dosage and time they receive the medication. In addition to that, sample details are also updated so that the laboratories can update the patient details once the results are available from the laboratory tests. The following diagram depicts the high-level implementation details of the treatment service:

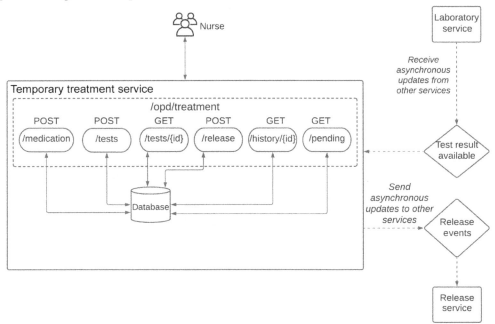

Figure 6.3 – Temporary treatment service

According to the preceding diagram, the service exposes a REST API to execute the tasks related to the temporary treatment and testing scope. The service has a REST API with `/opd/treatment` as the context path to showcase the functionality of this service. Under this API, there are six resources for executing specific functions within the service:

- A resource with the `/medication` context path with the `POST` method to update details on medication provided to the patient with the schedules.
- To report test results, there is a resource with the `/tests` context path with the `POST` method. All the details related to the tests are updated through this resource.
- To view the test reports for a particular patient, there is another resource under the `/tests/{id}` context path that is defined with the ID of the patient in the context path.
- To view the history of previous medications provided to the patient, there is a resource under the `/history/{id}` context path with the `GET` method.
- The resource under `/release` with the `POST` method is used to initiate the patient release process.
- There is another resource for viewing the pending treatments for patients who were inspected by the physicians but have not been attended to by the nursing staff. The resource with the `/pending` context path with the `GET` method is defined for this purpose.

For interservice communication, this service uses an asynchronous communication approach:

- Once the test results are available to be collected, the laboratory service will send an asynchronous notification to the service so that a person can be sent to collect the reports. In addition to that, if the digital reports are available, they will also be sent as an asynchronous event to this service.
- The patient release initiation events are published asynchronously to the release service.

The main consumers of this service would be nursing officers who take care of the treatments, medication, and collecting the samples.

Releasing the patient

The final stage of the OPD patient treatment process is to release the patient from the unit for further treatment by admitting them to the hospital or letting them return home. This process involves analyzing the patient's test results, treatment history, and input from the nurses. A physician will then decide on the patient's release procedure and inform the medical staff to execute it. The following diagram depicts the high-level architecture of the patient release service implementation:

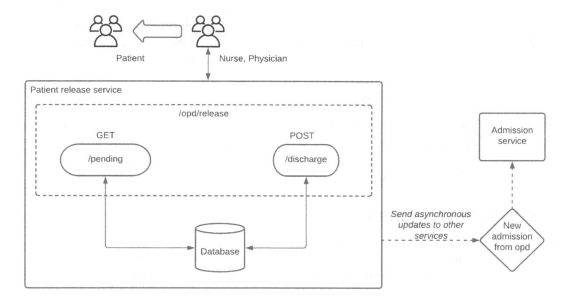

Figure 6.4 – Patient release service

The patient release service has an API that exposes the functionality related to the patient discharge procedure. A REST API is implemented with the `/opd/release` context path to showcase the functionality of the service. This API consists of two resources:

- A resource to view the pending release requests is implemented under the `/pending` context path with the GET method.

- Another resource under the `/discharge` context path is implemented with the POST method to update the patient record with the discharged state, as well as the required post-discharge medication that needs to be given to the patient.

This service communicates with other services using the asynchronous communication model. The details of the discharge event and admission event are shared with other services using an asynchronous communication model.

The consumers of this service are nurses and physicians who are involved in the patient release process.

Service composition

One major challenge with implementing microservices is handling data. As we discussed in the preceding section, each microservice has a database to store the data relevant to the service. But there can be many situations where these services need to share common data between them. As an example, an employee ID needs to be kept by all the services. At the same time, there can be many situations where the end user application needs to retrieve data from multiple services and show that data in an aggregated view. There are multiple approaches to implementing such use cases with API composition.

API composition layer

One approach to handling the operations that span multiple services is to use the API composition pattern, where there are intermediate services in front of the microservices to call different microservices and aggregate the responses to be displayed on the client side. We will discuss this option in the upcoming sections.

Client-side composition

The last resort in implementing operations that span multiple services is to implement them on the clientside, within the mobile or web application. This approach is also commonly used since the client machines are also becoming more and more powerful.

Security for services

Another key functional requirement of REST APIs is authentication and authorization. The services that we've discussed so far are consumed by client applications such as mobile or web apps, which are running on client devices such as computers, laptops, mobile phones, or tablet computers. Hence, protecting these services from malicious users is critical in this application since these applications deal with patient health information, which is a highly regulated information category, across many regions. Instead of implementing security at each service layer and repeating the code, we can use an external component to provide the required security without changing anything regarding the microservice's implementation. This functionality is provided by the API gateway layer, which we will discuss in detail in the next section.

Defining the solution architecture

Let's put all the knowledge we have gathered so far into a solution architecture so that we can start implementing the microservice-based OPD application. This solution architecture is a simplified version of the reference architecture that we discussed in *Chapter 5, Designing a Microservice Architecture with NATS*:

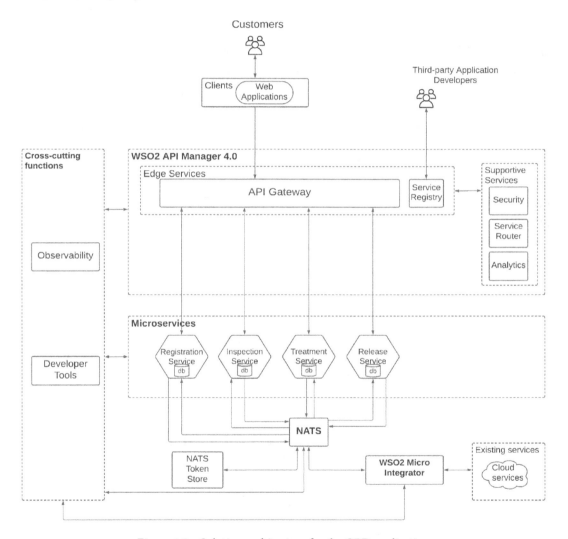

Figure 6.5 – Solution architecture for the OPD application

Let's take a deeper look at our solution architecture to understand the components we have used to build the solution.

Clients (Consumers)

Let's start from the top section of the preceding diagram, where we have the **Customers** or **Clients** who are accessing the OPD application. We have chosen a simple web application as the client to represent different types of API consumers that can exist in real-world applications. In addition to that, we will discuss how third-party application developers can utilize the overall solution to build more applications, such as our OPD application.

API management (with WSO2 API Manager 4.0)

The clients will access the services that are offered through the microservices via the API gateway. This gateway component utilizes other components such as security, service router, and analytics to apply various policies and monitoring on top of the API's usage. In addition to that, it also utilizes a service registry that will be used by the application developers to identify the available services in the system.

Microservices in Go

Then, we have the microservices layer, where we implement the services that we identified in the first section of this chapter. We are going to use the Go programming language to implement these services. Each service has a database to store the data related to that service. We will use MySQL as the database of choice for this application since it is open source and provides the structure that is required to store data related to patients.

NATS server

There are microservices that are built using NATS as the inter-service communication mechanism. The NATS servers are deployed as a cluster to support the availability and capacity requirements of the solution.

NATS token store

We will use OAuth2-based security for NATS servers to authenticate with the clients (microservices). These tokens are managed using the NATS token store, which is a separate file location that is independent of the other components.

Integration (with WSO2 Micro Integrator 4.0)

To showcase the anti-corruption layer capabilities, we will use WSO2 Micro Integrator 4.0, which is an open source integration platform that can integrate with existing services as well as the NATS messaging system.

Existing services

We will use a sample application that runs on the cloud to depict real-world, cloud-hosted applications that are used in enterprise solutions. This service will integrate with the microservices via the integration platform.

Cross-cutting functions

In addition to the core services, we will be using developer tools to develop microservices and supported artifacts on API management and integration components. Also, we will be using observability tools and mechanisms to observe the system to a certain extent within the scope of this example.

Components that were skipped

In this solution architecture, we have skipped some of the components that are present in our reference architecture for simplicity. The components that we have excluded from this sample implementation include the following:

- Delivery services (firewall, proxy, load balancer)
- Automation (CI/CD pipelines)
- Governance (microservices governance)
- Infrastructure provisioning

We had to keep those components away from this sample implementation due to the scope of this book. Those components need to be included when building enterprise-grade solutions with microservices.

In the next section, we'll understand what products and tools we are going to use in the deployment architecture.

Deployment architecture

In a deployment architecture diagram, we typically demonstrate the products and their deployment models, along with the interactions of these components, at a higher level. The following is our deployment architecture, which is built on top of the solution architecture diagram we discussed in the previous section:

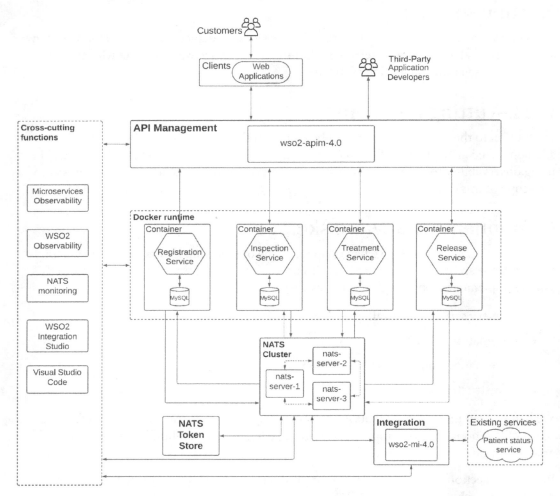

Figure 6.6 – Deployment architecture for the OPD application

Let's take a closer look at the preceding diagram by going through the details related to the deployment aspect of the solution.

WSO2 API Manager 4.0

We will use the open source API management solution known as **WSO2 API Manager 4.0** in this sample project. It will be deployed as an all-in-one component as a single node. It can be deployed in a highly available and distributed manner in a real-world deployment. But for this sample application, we decided to go ahead with an all-in-one single instance deployment.

Docker runtime for microservices

We will run our microservices written in Go on top of the Docker runtime so that you can get the experience of running the microservices in a truly cloud-native environment.

NATS cluster

We will deploy a three-node NATS cluster to serve the messages coming from the microservices. That allows us to showcase various failure handling scenarios of the NATS platform with the clustered deployment.

NATS token store

We will use the local filesystem of the server that we use to run the solution as the token store. This will be utilized by the NATS server cluster for authentication purposes with OAuth2.

WSO2 Micro Integrator 4.0

We will use a single instance of WSO2 MI 4.0 to showcase the integration capabilities of the solution with the existing cloud services of the platform.

Patient status service

A sample Node.js application running on GitHub will be used to demonstrate the functionality of an existing cloud service. This service will showcase the status of a patient who is admitted to the OPD.

Cross-cutting functions

There are specific tools that will be used to develop and monitor the deployment. Those tools are mentioned here:

- **Visual Studio Code**: This is used to develop microservices in Go and edit any other configuration files. You can use any IDE you prefer for this purpose.

- **WSO2 Integration Studio**: This tool is used to build the integrations that are used by WSO2 MI and WSO2 APIM.

- **WSO2 Observability**: These tools are used to observe the WSO2 APIM and WSO2 MI components.

- **NATS monitoring**: NATS provides several URLs to monitor the status of the server. In addition to that, it provides several CLI tools to monitor the status.

- **Microservices observability**: We will use Docker-provided tools to observe the behavior of the microservices.

At this point, we have all the tools and designs to build the solution. We will discuss the microservices implementation and how to test it with a NATS cluster later in this chapter. The other support components will be discussed in the next three chapters. Now, let's build the solution, starting with the microservices.

Implementing microservices

We have already defined the interfaces of the microservices in the previous section. Let's understand the data structures that we are going to use in their implementation. The following diagram depicts the data structures and their associations with each microservice:

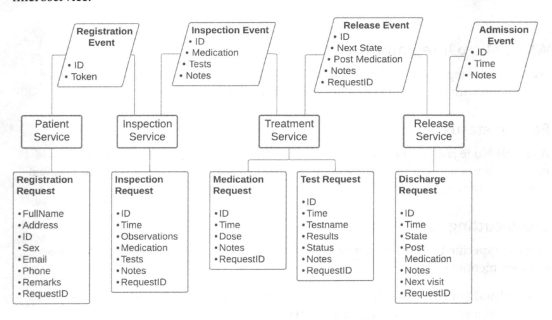

Figure 6.7 – Data structures and their associations with microservices

As depicted in the preceding diagram, each microservice is associated with its own data structures, as well as some common data structures that are used to share data via messages.

The patient service uses the patient registration process, so it uses the `RegistrationRequest` data type. The `RegistrationEvent` data type is used to communicate a patient registration to the inspection service.

The inspection service receives the registration event and uses that to process the inspection tasks. It uses the `InspectionRequest` data type to keep the inspection details. Once the inspection is done, it uses `InspectionEvent` to send the medication and testing instructions to the treatment service.

The treatment service receives `InspectionEvent` and starts executing the medication and testing tasks. It uses the `MedicationRequest` type to record medication activities and the `TestRequest` data type to record test activities. Once the patient is ready to be released from the unit, it shares that information via `ReleaseEvent` with the release service.

The release service receives `ReleaseEvent` and starts executing the patient discharge process. It uses the `DischargeRequest` data type to record the patient discharge activities. If the patient needs to be admitted to a hospital ward, it will share that information using `AdmissionEvent`.

The microservices interact with each other through the NATS messaging platform. The following diagram depicts the interactions that happen throughout the execution of the OPD application:

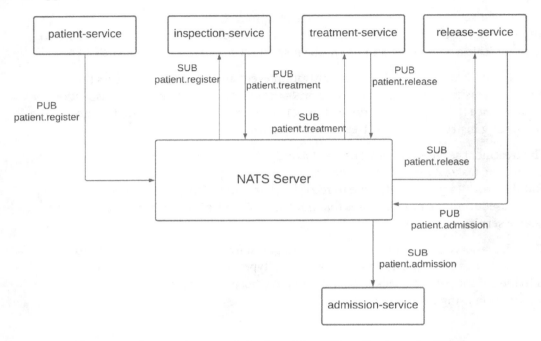

Figure 6.8 – Interservice communication of the OPD application using NATS

The preceding diagram depicts how each of the microservices shares information with other microservices using the NATS server. The patient service publishes the patient registration events via the `patient.register` subject name, which is subscribed by the inspection service. The inspection service publishes the patient treatment events via the `patient.treatment` subject, which is subscribed by the treatment service. The treatment service publishes events related to patient release via the `patient.release` subject, which is subscribed by the release service. If the patient needs to be admitted to hospital for long-term treatments, the release service publishes those events to the `patient.admission` subject, which is subscribed to by the external admission service.

The full source code of the implementation can be found at `https://github.com/PacktPublishing/Designing-Microservices-Platforms-with-NATS/tree/main/chapter6`.

Let's go through the code and understand the major concepts that we have used within the code. The code is organized in the following folder structure:

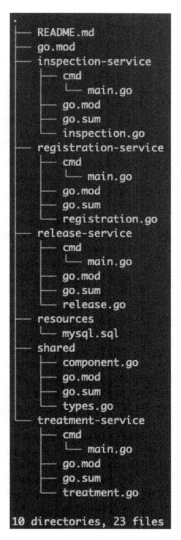

```
.
├── README.md
├── go.mod
├── inspection-service
│   ├── cmd
│   │   └── main.go
│   ├── go.mod
│   ├── go.sum
│   └── inspection.go
├── registration-service
│   ├── cmd
│   │   └── main.go
│   ├── go.mod
│   ├── go.sum
│   └── registration.go
├── release-service
│   ├── cmd
│   │   └── main.go
│   ├── go.mod
│   ├── go.sum
│   └── release.go
├── resources
│   └── mysql.sql
├── shared
│   ├── component.go
│   ├── go.mod
│   ├── go.sum
│   └── types.go
└── treatment-service
    ├── cmd
    │   └── main.go
    ├── go.mod
    ├── go.sum
    └── treatment.go

10 directories, 23 files
```

Figure 6.9 – Folder structure of the source code for the OPD application

The application is implemented in the Go programming language and each microservice is developed as a separate component that can be deployed independently. There are separate directories for each microservice, and two additional directories are created for the shared component, which is used by all the individual microservices as an external dependency and to store the SQL scripts to create the required databases for the application. Each microservice has a subdirectory that contains the main program that starts the relevant microservice. Let's go through the source code of each microservice.

Registration service

The main business logic of the registration service is implemented in the `registration.go` source file. It has separate methods to handle each of the API resource paths defined for the service. Let's go through the important code segments of the code. You can always refer to this book's GitHub repository for the full source code.

The following code segment is used to register the various resources with their implementation logic. It starts the HTTP server to serve the requests coming from external applications' North-South traffic:

```go
// ListenAndServe takes the network address and port that
// the HTTP server should bind to and starts it.
func (s *Server) ListenAndServe(addr string) error {
    r := mux.NewRouter()
    router := r.PathPrefix("/opd/patient/").Subrouter()
    // Handle base path requests
    // GET /opd/patient
    router.HandleFunc("/", s.HandleHomeLink)
    // Handle registration requests
    // POST /opd/patient/register
    router.HandleFunc("/register",
      s.HandleRegister).Methods("POST")
    // Handle update requests
    // PUT /opd/patient/update
    router.HandleFunc("/update", s.HandleUpdate)
      .Methods("PUT")
    // Handle view requests
    // GET /opd/patient/view/{id}
    router.HandleFunc("/view/{id}",
      s.HandleView).Methods("GET")
```

```go
// Handle token requests
// GET /opd/patient/token
router.HandleFunc("/token/{id}",
    s.HandleToken).Methods("GET")
// Handle token reset requests
// GET /opd/patient/token/reset/{id}
router.HandleFunc("/token/reset/{id}",
    s.HandleTokenReset).Methods("GET")
//router.HandleFunc("/events/{id}",
    deleteEvent).Methods("DELETE")
//log.Fatal(http.ListenAndServe(":8080", router))
l, err := net.Listen("tcp", addr)
if err != nil {
    return err
}
    srv := &http.Server{
        Addr: addr,
        Handler: router,
        ReadTimeout: 10 * time.Second,
        WriteTimeout: 10 * time.Second,
        MaxHeaderBytes: 1 << 20,
}
    go srv.Serve(l)
    return nil
}
```

These registered methods execute the business logic related to the resource. As an example, the following `HandleRegister` method takes care of registering patients by reading the input payload and inserting that information into the `patient_details` database. Once this information has been stored, it sends an event to the inspection service with the patient ID and the token number:

```go
// HandleRegister processes patient registration requests.
func (s *Server) HandleRegister(w http.ResponseWriter, r
    *http.Request) {
    body, err := ioutil.ReadAll(r.Body)
    if err != nil {
```

```go
        http.Error(w, "Bad Request", http.StatusBadRequest)
        return
}
var registration *shared.RegistrationRequest
err = json.Unmarshal(body, &registration)
if err != nil {
        http.Error(w, "Bad Request", http.StatusBadRequest)
        return
}
// Insert data to the database
db := s.DB()
insForm, err := db.Prepare("INSERT INTO
  patient_details(id, full_name, address, sex, phone,
    remarks) VALUES(?,?,?,?,?,?)")
if err != nil {
        panic(err.Error())
}
insForm.Exec(registration.ID, registration.FullName,
  registration.Address, registration.Sex,
    registration.Phone, registration.Remarks)
// Tag the request with an ID for tracing in the logs.
registration.RequestID = nuid.Next()
fmt.Println(registration)
// Publish event to the NATS server
nc := s.NATS()
//var registration_event shared.RegistrationEvent
tokenNo := generateTokenNumber(0)
registration_event :=
  shared.RegistrationEvent{registration.ID, tokenNo}
 reg_event, err := json.Marshal(registration_event)
if err != nil {
        log.Fatal(err)
        return
}
```

```
    log.Printf("requestID:%s - Publishing registration
      event with patientID %d\n", registration.RequestID,
        registration.ID)
    // Publishing the message to NATS Server
    nc.Publish("patient.register", reg_event)
    json.NewEncoder(w).Encode(registration_event)
}
```

There is a separate method for handling the patient information update requests. It reads the payload and updates the patient record by calling the database. The code for the update method is shown here:

```
// HandleUpdate processes requests to update patient details.
func (s *Server) HandleUpdate(w http.ResponseWriter, r
    *http.Request) {
    //patientID := mux.Vars(r)["id"]
    body, err := ioutil.ReadAll(r.Body)
    if err != nil {
        http.Error(w, "Bad Request", http.StatusBadRequest)
        return
    }
    var request *shared.RegistrationRequest
    err = json.Unmarshal(body, &request)
    if err != nil {
        http.Error(w, "Bad Request", http.StatusBadRequest)
        return
    }
     db := s.DB()
    insForm, err := db.Prepare("UPDATE patient_details SET
      full_name=?, address=?, sex=?, phone=?, remarks=?
        WHERE id=?")
    if err != nil {
        panic(err.Error())
    }
    insForm.Exec(request.FullName, request.Address,
      request.Sex, request.Phone, request.Remarks,
        request.ID)
```

```
      json.NewEncoder(w).Encode("Record for Patient updated
         successfully")
}
```

There is a separate method for viewing the patient records. HandleView reads the information from the database and returns the information to the caller. The source code for this method is mentioned here:

```
// HandleView processes requests to view patient data.
func (s *Server) HandleView(w http.ResponseWriter, r
    *http.Request) {
    patientID := mux.Vars(r)["id"]
    // Insert data to the database
    db := s.DB()
    selDB, err := db.Query("SELECT * FROM patient_details
      WHERE ID=?", patientID)
    if err != nil {
        panic(err.Error())
    }
    registration := shared.RegistrationRequest{}
    for selDB.Next() {
        var id, phone int
        var full_name, address, sex, remarks string
        err = selDB.Scan(&id, &full_name, &address, &sex,
            &phone, &remarks)
        if err != nil {
            panic(err.Error())
        }
        registration.ID = id
        registration.FullName = full_name
        registration.Address = address
        registration.Sex = sex
        registration.Phone = phone
        registration.Remarks = remarks
    }
        fmt.Println(registration)
```

```
        json.NewEncoder(w).Encode(registration)
}
```

At this point, we have an idea about how the registration service is implemented. To try out the functionality of the registration service, you need to execute the following steps:

1. Create a database called `opd_data` in MySQL Server and create the tables by executing the SQL script available in the `resources/mysql.sql` file.

2. Start `nats-server` with the default URL.

3. Go inside the `registration-service` directory.

4. Start the registration service by executing the following command. Make sure that you modify the parameter names for the database name, username, and password so that they match your database deployment:

```
$ go run cmd/main.go -dbName opd_data -dbUser username -
dbPassword password
```

5. Create a patient registration request with the following command:

```
$ curl "http://localhost:9090/opd/patient/register" -X
POST -d '{"full_name":"chanaka fernando","address":"44,
sw19, london","id":200, "sex":"male", "phone":222222222}'
```

You should get a response similar to the one shown here:

```
{"id":200,"token":1}
```

However, the OPD application is still not complete. We will go through the rest of the microservices in the upcoming sections.

Inspection service

This service listens to the registration events that have been published on the NATS server and executes the tasks related to patient inspection. The code is included inside the `inspection.go` source file. This implementation has a similar flow to the previous service, where we have a separate method to register the resources against the methods. Since we already looked at that code in the previous section, we won't be repeating that code here. You can refer to the source code of this service for more information. Instead, we'll go through a few important methods that execute the business logic. The first method is `ListenRegisterEvents`, which is used to listen to the registration events that are published by the registration service on the `patient.register` subject. Here, you can see the code related to subscribing to the NATS server. Once the event has been received, that information is stored in a database for further processing. This code is shown here:

```go
// Listen to patient registration events
func (s *Server) ListenRegisterEvents() error {
    nc := s.NATS()
    nc.Subscribe("patient.register", func(msg *nats.Msg) {
    var req *shared.RegistrationEvent
    err := json.Unmarshal(msg.Data, &req)
    if err != nil {
        log.Printf("Error: %v\n", err)
    }
    log.Printf("New Patient Registration Event received for
        PatientID %d with Token %d\n",req.ID, req.Token)
    // Insert data to the database
    db := s.DB()
    insForm, err := db.Prepare("INSERT INTO
        patient_registrations(id, token) VALUES(?,?)")
    if err != nil {
        panic(err.Error())
    }
    insForm.Exec(req.ID, req.Token)
    })
    return nil
}
```

Once the events have been received and stored, the users of the inspection service can view the pending inspections. This logic is implemented as a separate method. It simply reads from the database and responds with the pending inspections. In a real-world implementation, we could remove this intermediate database and directly write to the UI. However, that is outside the scope of this book. The following code segment showcases the logic of getting pending inspections from the database:

```
// HandlePending processes requests to view pending
inspections.
func (s *Server) HandlePending(w http.ResponseWriter, r
   *http.Request) {
   // Retrieve pending inspections from the database
   db := s.DB()
   selDB, err := db.Query("SELECT * FROM
      patient_registrations")
   if err != nil {
       panic(err.Error())
   }
       type allRegistrations []shared.RegistrationEvent
       var registrations = allRegistrations{}
   for selDB.Next() {
       var newRegistration shared.RegistrationEvent
       var id int
       var token uint64
       err = selDB.Scan(&id, &token)
       if err != nil {
           panic(err.Error())
       }
       newRegistration.ID = id
       newRegistration.Token = token
       registrations = append(registrations,
           newRegistration)
   }
   fmt.Println(registrations)
   json.NewEncoder(w).Encode(registrations)
}
```

Once the inspections have been made, that information needs to be updated in the system so that medication and tests can be initiated via the treatment service. The following code segment showcases recording the inspection results to the system. This method updates the database with the related information and removes the pending inspection event related to this patient from the relevant table. Finally, it sends an event to the treatment service with the information related to the medication and testing that's required for this patient. This implementation is shown in the following code segment:

```
// HandleRecord processes patient inspection record requests.
func (s *Server) HandleRecord(w http.ResponseWriter, r
    *http.Request) {
    body, err := ioutil.ReadAll(r.Body)
    if err != nil {
        http.Error(w, "Bad Request", http.StatusBadRequest)
        return
    }
    var inspection *shared.InspectionRequest
    err = json.Unmarshal(body, &inspection)
    if err != nil {
        http.Error(w, "Bad Request", http.StatusBadRequest)
        return
    }
    // Insert data to the database
    db := s.DB()
    insForm, err := db.Prepare("INSERT INTO
      inspection_details(id, time, observations,
        medication, tests, notes) VALUES(?,?,?,?,?,?)")
    if err != nil {
        panic(err.Error())
    }
    insForm.Exec(inspection.ID, inspection.Time,
      inspection.Observations, inspection.Medication,
        inspection.Tests, inspection.Notes)
    // Remove the entry from pending inspections table if
      it exists
    removeData, err := db.Prepare("DELETE FROM
      patient_registrations WHERE id=?")
```

```
if err != nil {
    panic(err.Error())
}
removeData.Exec(inspection.ID)
// Tag the request with an ID for tracing in the logs.
inspection.RequestID = nuid.Next()
fmt.Println(inspection)
// Publish event to the NATS server
nc := s.NATS()
inspection_event :=
    shared.InspectionEvent{inspection.ID, inspection.
      Medication,inspection.Tests, inspection.Notes}
reg_event, err := json.Marshal(inspection_event)
if err != nil {
    log.Fatal(err)
    return
}
log.Printf("requestID:%s - Publishing inspection event
    with patientID %d\n", inspection.RequestID,
      inspection.ID)
// Publishing the message to NATS Server
nc.Publish("patient.treatment", reg_event)
json.NewEncoder(w).Encode(inspection_event)
}
```

Once the inspection records have been updated, the history of the inspection records for a given patient may need to be viewed. For such use cases, we have implemented the following method, which will read from the inspection details database and provide the historical data for a given patient:

```
// HandleHistory processes requests to view inspection history.
func (s *Server) HandleHistory(w http.ResponseWriter, r
    *http.Request) {
patientID := mux.Vars(r)["id"]
// Read data from the database
db := s.DB()
selDB, err := db.Query("SELECT * FROM
```

```go
            inspection_details WHERE ID=?", patientID)
    if err != nil {
        panic(err.Error())
    }
    type allInspections []shared.InspectionRequest
    var inspections = allInspections{}
    for selDB.Next() {
      var newInspection shared.InspectionRequest
      var id int
      var time, observations, medication, tests, notes
        string
      err = selDB.Scan(&id, &time, &observations,
        &medication, &tests, &notes)
      if err != nil {
      panic(err.Error())
    }
      newInspection.ID = id
      newInspection.Time = time
      newInspection.Observations = observations
      newInspection.Medication = medication
      newInspection.Tests = tests
      newInspection.Notes = notes
      inspections = append(inspections, newInspection)
    }
    fmt.Println(inspections)
    json.NewEncoder(w).Encode(inspections)
}
```

To try out the functionality of the inspection service, you need to perform the following steps:

1. Go inside the inspection-service directory.

2. Start the inspection service by executing the following command. Make sure that you update the values for the database name, username, and password so that they match your database deployment:

```
$ go run cmd/main.go -dbName opd_data -dbUser username -
dbPassword password
```

3. Create a patient inspection request with the following command:

```
$ curl "http://localhost:9091/opd/inspection/record"
-X POST -d '{"id":200, "time":"2021/07/12 10:21 AM",
"observations":"ABC Syncrome", "medication":"XYZ x 3",
"tests":"FBT, UC", "notes":"possible Covid-19"}'
```

You should get a response similar to the following one:

```
{"id":200,"medication":"XYZ x 3","tests":"FBT,
UC","notes":"possible Covid-19"}
```

The next part of the application is the treatment service, which takes the inspection reports, provides treatments to the patient, and carries out various tests.

Treatment service

Inside the treatment service, we have implemented several methods for handling medication, tests, and release processes. We used the same approach as in the previous two services to register the resource paths with the respective methods. Hence, we are skipping that part of the code here. Instead, we'll go through the methods that have important business logic implemented.

The inspection events coming from the inspection service need to be received by the treatment service to start the treatments for the patients. This logic is implemented in the following code. It reads the events from the NATS server and stores that in a database for further processing:

```
// ListenTreatmentEvents listens to events coming from
inspection service
func (s *Server) ListenTreatmentEvents() error {
    nc := s.NATS()
    nc.Subscribe("patient.treatment", func(msg *nats.Msg) {
    var req *shared.InspectionEvent
    err := json.Unmarshal(msg.Data, &req)
    if err != nil {
        log.Printf("Error: %v\n", err)
```

```
    }
    log.Printf("New Patient Inspection Event received for
      PatientID %d\n",req.ID)
    // Insert data to the database
    db := s.DB()
    insForm, err := db.Prepare("INSERT INTO
      inspection_reports(id, medication, tests, notes)
        VALUES(?,?,?,?)")
    if err != nil {
        panic(err.Error())
    }
    insForm.Exec(req.ID, req.Medication, req.Tests,
      req.Notes)
    })
    return nil
}
```

Once this information has been stored in the database, the users of the treatment service, such as nurses, need to view the pending treatments of patients. This functionality is provided by the following code:

```
// HandlePendingView processes requests to view pending
treatments.
func (s *Server) HandlePendingView(w http.ResponseWriter, r
    *http.Request) {
    // Retrieve pending inspections from the database
    db := s.DB()
    selDB, err := db.Query("SELECT * FROM
      inspection_reports")
    if err != nil {
        panic(err.Error())
    }
    type allTreatments []shared.InspectionEvent
    var treatments = allTreatments{}
    for selDB.Next() {
      var newTreatment shared.InspectionEvent
      var id int
```

```
        var medication, tests, notes string
        err = selDB.Scan(&id, &medication, &tests, &notes)
        if err != nil {
            panic(err.Error())
        }
        newTreatment.ID = id
        newTreatment.Medication = medication
        newTreatment.Tests = tests
        newTreatment.Notes = notes
        treatments = append(treatments, newTreatment)
    }

    fmt.Println(treatments)
    json.NewEncoder(w).Encode(treatments)
}
```

The next step in the treatment process is to provide the medication that was recommended by the physician who did the inspection. The nurses will record this medication in the system by providing the relevant information. This functionality is executed in the following code segment. It will update the database table with the medication details and remove the entry from the pending medication table if this is the first medication for the patient:

```
// HandleMedicationRecord processes patient medication record
requests.
func (s *Server) HandleMedicationRecord(w
    http.ResponseWriter, r *http.Request) {
    body, err := ioutil.ReadAll(r.Body)
    if err != nil {
        http.Error(w, "Bad Request", http.StatusBadRequest)
        return
    }
    var medication *shared.MedicationRequest
    err = json.Unmarshal(body, &medication)
    if err != nil {
        http.Error(w, "Bad Request", http.StatusBadRequest)
        return
    }
    // Insert data to the database
```

```go
db := s.DB()
insForm, err := db.Prepare("INSERT INTO medication_
  reports(id, time, dose, notes) VALUES(?,?,?,?)")
if err != nil {
    panic(err.Error())
}
insForm.Exec(medication.ID, medication.Time,
  medication.Dose, medication.Notes)
// Remove the entry from pending medication table if it
exists
removeData, err := db.Prepare("DELETE FROM
  inspection_reports WHERE id=?")
if err != nil {
    panic(err.Error())
}
removeData.Exec(medication.ID)
json.NewEncoder(w).Encode("Record updated
  successfully")
}
```

A separate method is implemented to view the medication records of a particular patient. This information is useful for physicians when they're making decisions on the next stage of the patient (for example, releasing the patient or admitting them to the hospital). The source code for this method is shown here:

```go
// HandleHistoryView processes requests to view medication
history data.
func (s *Server) HandleHistoryView(w http.ResponseWriter, r
    *http.Request) {
    patientID := mux.Vars(r)["id"]
    // Select data from the database
    db := s.DB()
    selDB, err := db.Query("SELECT * FROM medication_
      reports WHERE ID=?", patientID)
    if err != nil {
        panic(err.Error())
    }
    type allMedications []shared.MedicationRequest
```

```
    var medications = allMedications{}
    for selDB.Next() {
      var newMedication shared.MedicationRequest
      var id int
      var time, dose, notes string
      err = selDB.Scan(&id, &time, &dose, &notes)
      if err != nil {
          panic(err.Error())
      }
      newMedication.ID = id
      newMedication.Time = time
      newMedication.Dose = dose
      newMedication.Notes = notes
      medications = append(medications, newMedication)
      }
      fmt.Println(medications)
      json.NewEncoder(w).Encode(medications)
}
```

A separate method is implemented to handle the tasks related to carrying out tests for patients. This includes use cases such as sample collection and updating the results. In either case, the following code segment reads the input data and updates the database with requests related to updating the test details for a patient:

```
// HandleTestRecord processes recording of tests related
requests.
func (s *Server) HandleTestRecord(w http.ResponseWriter, r
    *http.Request) {
    body, err := ioutil.ReadAll(r.Body)
    if err != nil {
        http.Error(w, "Bad Request", http.StatusBadRequest)
        return
    }
    var test *shared.TestRequest
    err = json.Unmarshal(body, &test)
    if err != nil {
        http.Error(w, "Bad Request", http.StatusBadRequest)
        return
```

```
    }
    // Insert data to the database
    db := s.DB()
     insForm, err := db.Prepare("INSERT INTO test_reports
        (id, time, test_name, results, status, notes)
          VALUES(?,?,?,?,?,?)")
    if err != nil {
        panic(err.Error())
    }
    insForm.Exec(test.ID, test.Time, test.TestName,
       test.Results, test.Status, test.Notes)
     json.NewEncoder(w).Encode("Test recorded
       successfully")
    }
```

To view the test results and various test-related information, we have implemented a
separate method. This method will read from the database and provide the historical
information of tests that have been carried out for a given patient. The source code for this
method is shown here:

```
// HandleTestView processes requests to view test data.
func (s *Server) HandleTestView(w http.ResponseWriter, r
    *http.Request) {
    patientID := mux.Vars(r)["id"]
    // Insert data to the database
    db := s.DB()
    selDB, err := db.Query("SELECT * FROM test_reports
      WHERE ID=?", patientID)
    if err != nil {
        panic(err.Error())
    }
    type allReports []shared.TestRequest
    var reports = allReports{}
    for selDB.Next() {
      var newReport shared.TestRequest
      var id int
      var time, test_name, results, status, notes string
```

```
    err = selDB.Scan(&id, &time, &test_name, &results,
        &status, &notes)
    if err != nil {
        panic(err.Error())
    }
    newReport.ID = id
    newReport.Time = time
    newReport.TestName = test_name
    newReport.Results = results
    newReport.Status = status
    newReport.Notes = notes
    reports = append(reports, newReport)
    }
    fmt.Println(reports)
    json.NewEncoder(w).Encode(reports)
    //defer db.Close()
}
```

Once the patient is done with the medication and tests, they need to be released from the OPD. This process is initiated by the physician based on the medication and test results. This initiation task is handled by the treatment service. It will send an event to the release service with information on post medication and the next steps. This functionality is implemented in the following code:

```
// HandleRelease processes requests to initiate a patient
release.
func (s *Server) HandleRelease(w http.ResponseWriter, r
    *http.Request) {
    body, err := ioutil.ReadAll(r.Body)
    if err != nil {
        http.Error(w, "Bad Request", http.StatusBadRequest)
        return
    }
    var release *shared.ReleaseEvent
    err = json.Unmarshal(body, &release)
    if err != nil {
        http.Error(w, "Bad Request", http.StatusBadRequest)
        return
```

```
    }
    // Tag the request with an ID for tracing in the logs.
    release.RequestID = nuid.Next()
    fmt.Println(release)
    // Publish event to the NATS server
    nc := s.NATS()
    release.RequestID = nuid.Next()
    release_event := shared.ReleaseEvent{release.ID,
      release.Time, release.NextState, release
        .PostMedication, release.Notes, release.RequestID}
    rel_event, err := json.Marshal(release_event)
    if err != nil {
        log.Fatal(err)
        return
    }
    log.Printf("requestID:%s - Publishing inspection event
      with patientID %d\n", release.RequestID, release.ID)
    // Publishing the message to NATS Server
    nc.Publish("patient.release", rel_event)
    json.NewEncoder(w).Encode("Release event published")
}
```

Now that we have an understanding of the treatment service implementation, let's try out the functionality of the service:

1. Go inside the `treatment-service` directory.

2. Start the treatment service by executing the following command. Make sure that you update the values for the database name, username, and password parameters so that they match your database deployment:

```
$ go run cmd/main.go -dbName opd_data -dbUser username -
dbPassword password
```

3. Create a medication record request with the following command:

```
$ curl "http://localhost:9092/opd/treatment/medication"
-X POST -d '{"id":200,"time":"2021 07 12 4:35
PM","dose":"xyz x 1, abc x 2","notes":"low fever"}'
```

4. You should get a response similar to the following one:

```
"Record updated successfully"
```

You can try out other resources by following the README.md file in this book's GitHub repository. For the moment, let's skip the testing and move on to the next microservice.

Release service

This is the final part of the OPD application and is where patients are released from the OPD unit after their initial inspection and treatment. In this microservice, we implement the methods that support the patient release process. The first method is implemented to listen to the release events that are published by the treatment service. These events listen through the NATS server with patient.release. Once the event has been received, it is stored in the database for further processing. These implementation details are mentioned in the following code segment:

```
// ListenReleaseEvents Listen to release events and update the
temporary table
func (s *Server) ListenReleaseEvents() error {
    nc := s.NATS()
    nc.Subscribe("patient.release", func(msg *nats.Msg) {
    var req *shared.ReleaseEvent
    err := json.Unmarshal(msg.Data, &req)
    if err != nil {
        log.Printf("Error: %v\n", err)
    }
    log.Printf("New Patient Release Event received for
      PatientID %d\n",req.ID)
    // Insert data to the database
    db := s.DB()
    insForm, err := db.Prepare("INSERT INTO release_reports
      (id, time, next_state, post_medication, notes)
        VALUES(?,?,?,?,?)")
    if err != nil {
        panic(err.Error())
    }
    insForm.Exec(req.ID, req.Time, req.NextState,
      req.PostMedication, req.Notes)
```

```
    })
    return nil
}
```

Once the medical staff member is ready to process another patient release task, that person will check out the pending releases and execute the release process. This functionality is implemented through the following method, which will read the pending releases from the database and update the consumer with that information:

```
// HandlePendingView processes requests to view pending
releases.
func (s *Server) HandlePendingView(w http.ResponseWriter, r
    *http.Request) {
    // Retrieve pending inspections from the database
    db := s.DB()
    selDB, err := db.Query("SELECT * FROM release_reports")
    if err != nil {
        panic(err.Error())
    }
    type allReleases []shared.ReleaseEvent
    var releases = allReleases{}
    for selDB.Next() {
      var newRelease shared.ReleaseEvent
      var id int
      var time, next_state, post_medication, notes string
      err = selDB.Scan(&id, &time, &next_state,
        &post_medication, &notes)
      if err != nil {
        panic(err.Error())
      }
      newRelease.ID = id
      newRelease.Time = time
      newRelease.NextState = next_state
      newRelease.PostMedication = post_medication
      newRelease.Notes = notes
      releases = append(releases, newRelease)
    }
    fmt.Println(releases)
```

```
        json.NewEncoder(w).Encode(releases)
}
```

The patient discharge process verifies the patient information, provides the necessary instructions, and posts medications to the patient if any have been recommended by the physician. In addition to that, if there is a need for the patient to come back to the hospital, that information also needs to be updated in the system. If the patient needs to be admitted to the hospital for further treatment, that information should also be sent to the admission service. These tasks are implemented within the following code segment:

```
// HandleDischargeRecord processes patient discharge requests.
func (s *Server) HandleDischargeRecord(w
    http.ResponseWriter, r *http.Request) {
    body, err := ioutil.ReadAll(r.Body)
    if err != nil {
        http.Error(w, "Bad Request", http.StatusBadRequest)
        return
    }
    var discharge *shared.DischargeRequest
    err = json.Unmarshal(body, &discharge)
    if err != nil {
        http.Error(w, "Bad Request", http.StatusBadRequest)
        return
    }
    // Insert data to the database
    db := s.DB()
    insForm, err := db.Prepare("INSERT INTO
      discharge_details(id, time, state, post_medication,
        notes, next_visit) VALUES(?,?,?,?,?,?)")
    if err != nil {
        panic(err.Error())
    }
     insForm.Exec(discharge.ID, discharge.Time,
        discharge.State, discharge.PostMedication,
          discharge.Notes, discharge.NextVisit)
    // Remove the entry from pending release table if it
      exists
    removeData, err := db.Prepare("DELETE FROM
```

```go
        release_reports WHERE id=?")
    if err != nil {
        panic(err.Error())
    }
    removeData.Exec(discharge.ID)
    // Send admission request if required
    if discharge.State == "admission" {
      discharge.RequestID = nuid.Next()
      // Publish event to the NATS server
      nc := s.NATS()
      //var registration_event shared.RegistrationEvent
      admission_event := shared.AdmissionEvent
        {discharge.ID, discharge.Time, discharge.Notes}
      reg_event, err := json.Marshal(admission_event)
      if err != nil {
          log.Fatal(err)
          return
      }
      log.Printf("requestID:%s - Publishing inspection
        event with patientID %d\n", discharge.RequestID,
          discharge.ID)
      // Publishing the message to NATS Server
      nc.Publish("patient.admission", reg_event)
    }
    json.NewEncoder(w).Encode("Patient discharge recorded
      successfully")
}
```

Now that we have an understanding of the treatment release service implementation, let's try out the functionality of the service:

1. Go inside the release-service directory.

2. Start the release service by executing the following command. Make sure that you update the values for the database name, username, and password so that they match your database deployment:

```
$ go run cmd/main.go -dbName opd_data -dbUser username -
dbPassword password
```

3. Create a discharge request with the following command:

```
$ curl "http://localhost:9093/opd/release/discharge"
-X POST -d '{"id":200,"time":"2021 07 12 9:35
PM","state":"discharge","post_medication":"NM x 10
days","next_visit":"2021 08 12 09:00AM"}'
```

You should get a response similar to the following one:

```
"Patient discharge recorded successfully"
```

There is another directory that you find in the source code called `shared` that contains the implementation that is common across all the microservices. Hence, we have implemented that as a separate component in Go and used it in the microservice implementation as a shared library, similar to any other shared libraries, such as `nats.io`.

You can try out other resources by following the `README.md` file in this book's GitHub repository. For the moment, let's skip the testing and move on to the next section, where we will set up the NATS server cluster and try out the full application.

Setting up the NATS server cluster

NATS servers can be clustered to support high-volume systems and to provide better availability. NATS uses a simple clustering protocol to connect with other servers via gossiping and connecting to all the servers that a particular server is aware of. Once clients connect to a given server, the clients are informed about the current cluster members. We discussed NATS clustering in detail in *Chapter 3, What Is NATS Messaging?*. For this chapter, let's create a three-node cluster to try out our OPD microservice application.

Starting up the three-node NATS server cluster

Let's start a three-node cluster by specifying the client port and the cluster port, as shown here:

1. Start the first server as the **seed server**:

```
$ nats-server -p 4222 -cluster nats://localhost:4248 -D
```

2. Start the second server by specifying the cluster URL of the first (seed) server:

```
$ nats-server -p 5222 -cluster nats://localhost:5248
-routes nats://localhost:4248 -D
```

3. Start the third server by specifying the cluster URLs of the first (seed) server:

```
$ nats-server -p 6222 -cluster nats://localhost:6248
-routes nats://localhost:4248 -D
```

In the aforementioned commands, we used the `routes` option to specify the connection to the seed server. This server acts as the starting point for server discovery by other members of the cluster and the clients. When the second and third servers are starting up, you should see messages in the command window that reflect the servers connecting over the clustering protocol.

Assuming that all the NATS servers started without any error, our testing environment is ready with the NATS server cluster and the microservices implemented in Go. The next step is to get our hands dirty with the testing.

Trying out the sample application

At this point, we have implemented a basic OPD application using a microservice-based approach and have a NATS server cluster set up on our local computer. Now, let's go ahead and try out the sample application to verify its concepts, design, and implementation. The following diagram depicts the current state of the application:

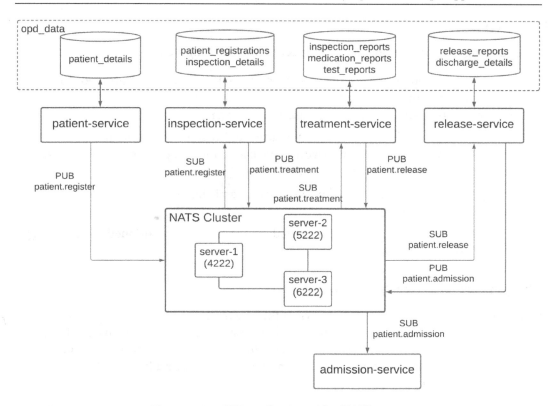

Figure 6.10 – OPD application with a NATS server

As depicted in the preceding diagram, individual microservices have separate databases to store the data related to the respective microservice. Also, each microservice uses the NATS messaging platform to share data between other microservices.

Let's go ahead and try out the sample application and its functions. We will use curl as the client to connect with the microservices. You can use any HTTP client here, such as **Postman** or **JMeter**.

Prerequisites

Before running the application, you must execute the following steps:

1. Download the source code of the application from this book's GitHub repository (`https://github.com/chanakaudaya/Designing-Microservices-Platforms-with-NATS/tree/main/chapter6`).

2. Create a database called `opd_data` in your local MySQL server and grant permissions for a user with a password. This username and password will be used when you're starting the applications.

3. Create tables that are mentioned in the `mysql.sql` database script, under the `/resources` directory.

4. Make sure that the NATS cluster is started appropriately, as mentioned in the *Setting up the NATS server cluster* section.

Starting the microservices

Let's go ahead and start the four microservices that are included in the downloaded source code repository. All the commands mentioned here assume that you have navigated to the `chapter6` directory. Use a new terminal window to start each microservice mentioned here and make sure that you replace the `username` and `password` fields with the relevant details of your MySQL Server:

1. Start the registration microservice with the following commands. We will provide the NATS server URL of the first server when starting this service:

    ```
    $ cd registration-service
    $ go run cmd/main.go -dbName opd_data -dbUser username
    -dbPassword password -nats nats://localhost:4222
    ```

    Once the service has started properly, you should see a message similar to the following:

    ```
    2021/07/11 13:56:12 Starting NATS Microservices OPD
    Sample - Registration Service version 0.1.0
    2021/07/11 13:56:12 Listening for HTTP requests on
    0.0.0.0:9090
    ```

2. Start the inspection microservice with the following commands. We will provide the URL of the second NATS server when starting this service:

```
$ cd inspection-service
$ go run cmd/main.go -dbName opd_data -dbUser username
-dbPassword password -nats nats://localhost:5222
```

Once the service has started properly, you should see a message similar to the following:

```
2021/07/11 13:56:18 Starting NATS Microservices OPD
Sample - Inspection Service version 0.1.0
2021/07/11 13:56:18 Listening for HTTP requests on
0.0.0.0:9091
```

3. Start the treatment microservice with the following commands. We will provide the URL of the third NATS server when starting this service:

```
$ cd treatment-service
$ go run cmd/main.go -dbName opd_data -dbUser username
-dbPassword password -nats nats://localhost:6222
```

Once the service has started properly, you should see a message similar to the following:

```
2021/07/11 13:56:26 Starting NATS Microservices OPD
Sample - Treatment Service version 0.1.0
2021/07/11 13:56:26 Listening for HTTP requests on
0.0.0.0:9092
```

4. Finally, start the release microservice with the following commands. We will provide the URL of all three NATS servers when starting this service:

```
$ cd release-service
$ go run cmd/main.go -dbName opd_data -dbUser username
-dbPassword password -nats nats://localhost:4222, nats://
localhost:5222, nats://localhost:6222
```

Once the service has started properly, you should see a message similar to the following:

```
2021/07/11 13:54:25 Starting NATS Microservices OPD
Sample - Release Service version 0.1.0
2021/07/11 13:54:25 Listening for HTTP requests on
0.0.0.0:9093
```

Testing the OPD application

Now that we have the basic building blocks of a microservice-based application that communicates over NATS for interservice communication, let's take a patient through the OPD application.

Registering a patient

The first part of the OPD application is registering the patient in the hospital database. In this scenario, the patient registration information is stored in a database that is owned by the registration microservice. We can execute the following command to register a patient with an ID of 200. This ID is used as the primary key of the patient database:

```
curl "http://localhost:9090/opd/patient/register" -X POST -d
'{"full_name":"John Doe","address":"44, sw19, london","id":200,
"sex":"male", "phone":222222222}
```

You should get a response containing the ID of the patient and the token number that's been issued for the patient's inspection. A sample response is shown here:

```
{"id":200,"token":1}
```

Once the patient has registered at the reception and been issued with a token number, the patient goes to the inspection unit. The inspection unit already knows the patient's ID and token number since they were updated through an event to the inspection service.

Inspecting a patient

The physician will check the pending inspections by sending a request to the following URL of the inspection service:

```
curl "http://localhost:9091/opd/inspection/pending"
```

We will get a list of users pending inspection. At the moment, we only have one patient in the queue. You should get a response similar to the following:

```
{"id":200,"token":1}
```

The next step is to inspect the patient and record the inspection details on the system against the patient ID. To do that, we can execute the following command:

```
curl "http://localhost:9091/opd/inspection/record" -X POST -d
'{"id":200, "time":"2021/07/12 10:21 AM", "observations":"ABC
Syncrome", "medication":"XYZ x 3", "tests":"FBT, UC",
"notes":"possible Covid-19"}'
```

You should get a response similar to the following from the preceding request:

```
{"id":200,"medication":"XYZ x 3","tests":"FBT,
UC","notes":"possible Covid-19"}
```

Once the inspection is done, it will automatically send an event to the treatment service with the details of the inspection. This will let them start the medication process when the patient goes to the treatment section.

Providing treatments to a patient

The first thing nurses will do before treating a patient is view the pending treatment requests. We can view the pending treatment requests by executing the following command:

```
curl "http://localhost:9092/opd/treatment/pending"
```

This would provide a response containing pending treatments, similar to this:

```
[{"id":200,"medication":"XYZ x 3","tests":"FBT,
UC","notes":"possible Covid-19"}]
```

After reviewing this information, the nurses will start executing the medication process and do the necessary tests recommended by the physician via the inspection report. These activities should also be reported in the system.

We can use the following command to update a medication record for a patient:

```
curl "http://localhost:9092/opd/treatment/medication" -X POST
-d '{"id":200,"time":"2021 07 12 4:35 PM","dose":"xyz x 1, abc
x 2","notes":"low fever"}'
```

The aforementioned command will respond like so:

```
"Record updated successfully"
```

We can use the following command to update information about a test that is carried out on a patient:

```
curl "http://localhost:9092/opd/treatment/tests" -X
POST -d '{"id":200,"time":"2021 07 12 4:35 PM","test_
name":"FBC","status":"sample collected", "notes":"possible
covid-19"}'
```

The aforementioned command will respond like so:

```
"Test recorded successfully"
```

Once the patient is done with their treatments and testing, they will be released from the OPD unit by either being sent back home or being admitted to a hospital ward for long-term treatment. This process can be initiated with the following command:

```
curl "http://localhost:9092/opd/treatment/release" -X
POST -d '{"id":200,"time":"2021 07 12 8:35 PM","next_
state":"discharge","post_medication":"NM x 10 days"}'
```

This command will produce an output similar to the one shown here:

```
"Release event published"
```

Releasing a patient from OPD

The final phase of the OPD application is releasing the patient from the OPD unit. The first thing nurses will check for in the system is releases pending. We can get the pending releases by executing the following command:

```
curl "http://localhost:9093/opd/release/pending"
```

The aforementioned command will produce an output similar to the following:

```
[{"id":200,"time":"2021 07 12 8:35 PM","next_
state":"discharge","post_medication":"NM x 10 days"}]
```

After reviewing the details, the patient will be released from the OPD unit, and the details of the release need to be updated in the system. We can update their release details using the following command:

```
curl "http://localhost:9093/opd/release/discharge"
-X POST -d '{"id":200,"time":"2021 07 12 9:35
PM","state":"discharge","post_medication":"NM x 10 days","next_
visit":"2021 08 12 09:00AM"}'
```

You will get a response similar to this one:

```
"Patient discharge recorded successfully"
```

When you are testing the application, you can shut down one or two NATS servers. You will see that the application will function without any errors so long as there is one server available in the system.

With this, we have finished implementing a microservice-based application with NATS as the messaging layer for interservice communication.

Summary

In this chapter, we defined a use case for building an application that supports operations within an OPD of a hospital. Then, we discussed the solution architecture and deployment architecture of a microservice-based OPD application that uses NATS as the messaging platform. We identified the main components of the application and defined the data structures and interactions of microservices with each other and with external consumers. After that, we implemented a fully functioning OPD application using a microservice-based approach with the Go programming language. Then, we went through the code and discussed the implementation details using code segments from the application. After that, we discussed setting up a 3-node NATS cluster. Finally, we tried out the OPD application by connecting microservices with the NATS cluster using a real-world example of a patient going through the OPD process.

We will discuss the details of outer architecture components that we skipped in this chapter due to scope limitations in *Chapter 7*, *Securing a Microservices Architecture with NATS*, *Chapter 8*, *Observability with NATS in a Microservices Architecture*, and *Chapter 9*, *How Microservices and NATS Coexist with Integration Platforms*. We will talk about the security of microservices for North-South traffic as well as for East-West traffic, which involves the NATS server, in the next chapter.

7
Securing a Microservices Architecture with NATS

Security is no longer an afterthought in information systems design—it is a fundamental requirement of any system that we design today. In a microservices architecture, security becomes increasingly important since it exposes a much larger surface area to consumers (both genuine users and non-genuine users) due to the increased number of independent services that are deployed as microservices. Some traditional security mechanisms used in the enterprise world will not work well in a microservices context. We will discuss the security of microservices by considering the following two traffic flow patterns we have looked at in previous chapters:

- Security of North-South traffic
- Security of East-West traffic

In *Chapter 6, A Practical Example of Microservices with NATS*, we briefly discussed how an **application programming interface** (**API**) gateway can provide security features to the microservices for North-South traffic. In this chapter, we will discuss that topic in detail. In addition to that, we also discuss the security of East-West traffic using the security capabilities offered by the **Neural Autonomic Transport System** (**NATS**) messaging system. These are the primary areas that we are going to discuss in this chapter:

- Understanding security in the context of a microservices architecture
- Securing external communication
- Securing **inter-service communication** (**ISC**)
- Using NATS to secure ISC

By the end of this chapter, you will understand the concept of security in a microservices context and how to implement that with the usage of NATS messaging.

Technical requirements

In this chapter, we will be configuring the NATS server to enable security for client-server and server-server communication. In addition to that, we will implement a few sample applications in the Go programming language to try out client-server communication with security enabled. The following software components need to be installed on your computer to try out the examples mentioned in this chapter:

- Installation of the NATS server
- Installation of the Go programming language

The full source code of the examples used in this chapter can be found at `https://github.com/PacktPublishing/Designing-Microservices-Platforms-with-NATS/tree/main/chapter7`.

Understanding security in the context of a microservices architecture

Microservices architectures encourage decomposing an application into small (in scope), autonomous units that can be managed and deployed independently. If we compare this with a traditional monolithic application, one major difference is that a microservices architecture opens the security of the platform to a wider surface area. In a monolithic application, most of the communication happens within the application itself inside the same server and runtime, hence it does not require any advanced security for internal communication. But in the world of microservices, we need to secure the communication coming into the services (North-South traffic) as well as within the services (East-West traffic). The following diagram depicts this concept of two types of security that need to be handled in a microservices architecture:

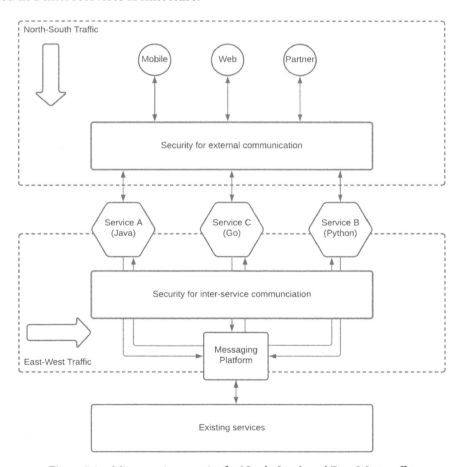

Figure 7.1 – Microservice security for North-South and East-West traffic

The preceding diagram depicts the two major interactions that we should secure within a microservices architecture. The first line of defense is the security of microservices for external consumers. These are the direct users of microservices, such as mobile, web, or partner applications. There are many established security mechanisms available in the industry for securing North-South traffic and East-West traffic, even before the emergence of microservices architectures. Let's briefly look at these general mechanisms offering security for services.

There are two primary areas in terms of security for services and applications regardless of the architecture (monolithic or microservices), as outlined here:

- **Authentication**: Verifies who the user is
- **Authorization**: Verifies what the user can do

There have been many different mechanisms used by enterprise applications to implement security for services to cover the authentication and authorization requirements. Some common mechanisms used for authentication include the following:

- **Basic authentication**: Using a username and password to authenticate the user
- **Token-based authentication**: Using a token (encrypted value) to authenticate the user
- **Certificate-based authentication**: Using a certificate to authenticate the user

On top of this, there are several methods used to implement authorization for services, including the following:

- **Role-based access control** (**RBAC**): Controls access to services based on roles and permissions
- **Attribute-based access control** (**ABAC**): Controls access to services based on attributes of the user
- **Delegated authorization**: A mechanism where a user delegates the right to access a service to a third-party application using an access grant

These different mechanisms are still used in many enterprise applications and services to implement security.

Security with OAuth 2.0

Open Authorization 2.0 (OAuth 2.0) is a modern approach that is evolving as the de facto standard in securing services. It is an authorization framework to build secure services by providing a simple-to-use approach for different client applications, including mobile, web, and desktop applications. It allows a client application (for example, mobile or web) to access a secured service on behalf of an authorized user with the consent of the user. One thing to note here is that OAuth 2.0 is not an authentication protocol; it is an authorization framework. The actual authentication may still happen using the general security mechanisms we discussed previously.

As an example, the user will be authenticated to the system by providing a username and password at the authentication step. Once that is done, the OAuth 2.0 framework allows the client application to generate a token based on that authentication step so that it can perform certain operations using that token on behalf of the user. The following diagram depicts a typical authorization flow within the OAuth 2.0 framework:

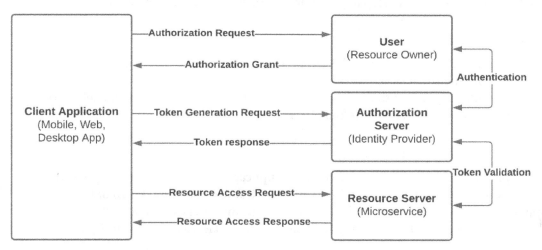

Figure 7.2 – OAuth 2.0 basic protocol flow

The OAuth 2.0 framework is built on top of four main components (roles). Those components and their interactions during a typical application or microservice security use case are depicted in the preceding diagram. Let's briefly define the roles first.

User (resource owner)

This is the end user who has the authority to access the resource. In the context of microservices, this is a valid user who can consume the protected microservice. Fine-grained access control can be implemented by defining user groups with specific permissions and then creating **scopes** that consist of one or more groups.

Client application

Users will access the system using a client application that is running on their computer or mobile phone. It can be a mobile application, a web application, or a desktop application.

Authorization server

This is the component where users are authenticated. Typically, this is an **identity provider** (**IdP**) solution that verifies the user's identity and responds with a **grant** that is used by the client in a later step to get an access token to access the resource.

Resource server

This is the actual server that runs the resource or the microservice. It receives access requests from the client, along with an access token. Then, the resource server will validate the access token by calling the authorization server (in the case of an opaque token) or by itself (in the case of **JSON Web Token** (**JWT**)).

OAuth 2.0 protocol flow

In a typical scenario, the user starts the client application by clicking on the application icon (mobile, desktop) or typing in the **Uniform Resource Locator** (**URL**) of the web application. If the application backend (for example, microservice) is protected with OAuth2.0 for security, the following set of actions take place, as depicted in *Figure 7.2*:

1. The client requests authorization from the user by redirecting the user to the authentication endpoint, which typically runs on the authorization server.

2. The user provides the credentials, any other details, and consent to the authorization server, and the server will respond to the client with an authorization grant.

3. The client will use this authorization grant (for example, an authorization code) to call the token endpoint in the authorization server and request an access token.

4. The authorization server validates the grant and provides the client with an access token.

5. The client sends a request to the resource server (for example, a microservice) with the access token to get access to the service.

6. The resource server validates the access token and if it is valid, responds with the requested details from the service.

Even though OAuth 2.0 is a topic of its own, this section provided a high-level introduction to how it works in general. More details on OAuth 2.0 can be found at `https://oauth.net/2/`. In the context of this book, we will be using OAuth 2.0 as the standard for securing microservices, but let's first discuss how to secure microservices for external access (North-South traffic).

Securing external communication

As we discussed in the previous section, OAuth2.0 is becoming the standard when it comes to securing APIs and microservices. In this section, we will see how we can implement OAuth 2.0-based security for microservices, with a few options that are available.

Implementing security at each microservice level

Given that microservices are developed by autonomous teams, they have the freedom to select the best technology stack for their respective microservices. At the same time, we have discussed that certain things need to be adhered to by all the microservices teams for better governance. Security is one such aspect that different teams need to agree on. It would help the clients of these services to follow a common, standards-based approach to consume these services.

Once the teams have agreed upon a certain approach—let's say, to use OAuth 2.0 as the security protocol to implement microservices—the next step is to implement it. Since microservices teams can use different programming languages to develop microservices, one possibility is to implement security at each microservice level according to the tools provided by the respective programming languages. Given that OAuth 2.0 is a standard and most languages do have implementations to easily support that, this will not be an issue for microservices teams. The following diagram depicts this concept:

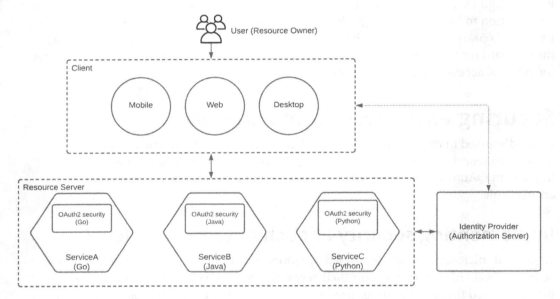

Figure 7.3 – Microservice security at each service layer

Figure 7.3 depicts the idea of implementing security at individual microservices' layers with the respective programming languages. A separate authorization server (**IdP**) can be used as the shared component that takes care of user and token management. Individual microservices can validate tokens with or without the help of the IdP, depending on the token type. Further details are provided here:

- **Self-contained JWT**: If the token sent by the client is a JWT, then the microservice can validate the token without calling the authorization server.

- **Opaque (reference) token**: In the case of an opaque token, the microservice must call the authorization server to validate the token since it does not contain any information about the client or the user.

One disadvantage of this approach is that each team must implement the same functionality repeatedly.

Implementing security using a sidecar

Another approach to implementing security for North-South traffic is using so-called **sidecar** models. A sidecar in a microservices context is a component that is attached to the main microservice and deployed along with that microservice. This approach is depicted here:

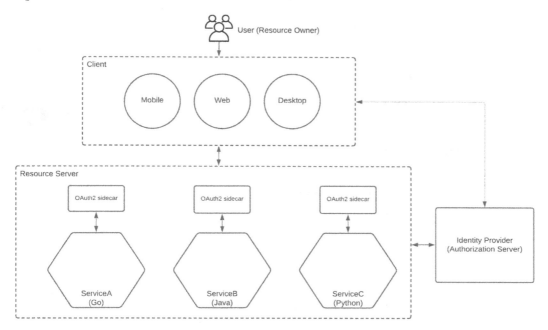

Figure 7.4 – Microservice security with a sidecar

As shown in the preceding diagram, each microservice has an OAuth 2.0 sidecar deployed along with the service. In the case of a containerized deployment, this sidecar component will run in the same container or same Pod (in Kubernetes) and has minimal network-level separation from the original microservice. There are certain open source technologies such as **Istio** and **Open Policy Agent (OPA)** that support this kind of security implementation for microservices. Another approach to implementing this model is to use API microgateways as sidecars to provide security for microservices. In this approach also, a separate resource server can be used to manage users and tokens.

An advantage of this model is that microservices do not need to implement security-related code at an individual microservices level. Instead, they can use this common implementation, which will run alongside the microservice as a separate runtime. The configuration of these sidecars is also done by the respective solutions (for example, Istio; OPA; microgateway).

Implementing security using an API gateway

The next possible approach to implement security for North-South traffic is to use an **API gateway**. In this approach, the microservices offload the security responsibility to a separate component that will run as a centralized component managed separately from the microservices. The following diagram depicts this concept:

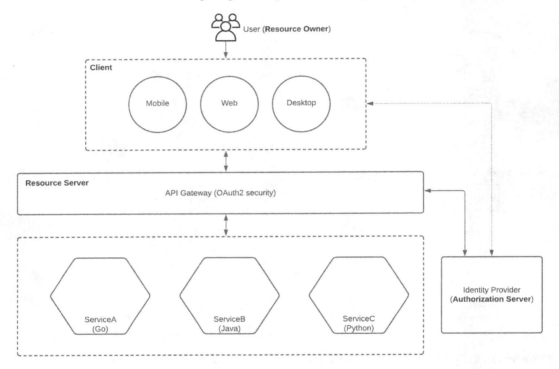

Figure 7.5 – Microservice security with an API gateway

As shown in the preceding diagram, in this model, microservices do not need to care about the security aspect, and the API gateway acts as the resource server that validates the OAuth 2.0 tokens. If the token is valid, depending on the token type, it will route the requests directly to the microservices. These microservices can implement **Secure Sockets Layer (SSL)** security for the sake of **Transport Layer Security (TLS)**, but do not need to worry about **Messaging-Level Security (MLS)**. This approach eliminates the need for implementing security at an individual microservice level.

We will be using this approach to implement OAuth 2.0 security for North-South traffic in this book. The next step in securing a microservices architecture is to secure the inner architecture where ISC happens. Let's discuss that in detail in the following section.

Securing ISC

The security of ISC (East-West within a microservices architecture needs to complement the security that is applied to North-South communication. One major difference between the security of North-South communication and the security of East-West communication is that the former deals with external systems and applications while the latter deals with internal systems. That allows us to consider options for ISC other than North-South communication.

Given that we use the NATS messaging platform as the intermediate component for ISC, the security implementation will depend on the security capabilities offered by NATS. We discussed the security features available in NATS in *Chapter 3, What Is NATS Messaging?*, and *Chapter 4, How to Use NATS in a Microservices Architecture?*, in detail. Let's first discuss how those features fit into the overall microservices security aspect and then implement a couple of options with a few examples using our **Out Patient Department (OPD)** application. Given that the security of ISC depends on the communication of the microservices with the NATS server, the following options are available for securing ISC:

- Using TLS
- Using **Application-Level Security (ALS)**

   a) Using authentication

   b) Using authorization

Let's discuss these options in detail in the following section.

Security with TLS

Since microservices run inside a secured network layer such as a **Militarized Zone (MZ)**, it is possible to implement ISC using the TLS protocol. In this approach, there is no additional security implemented when the microservices communicate with the NATS server other than the TLS protocol. This approach is depicted here:

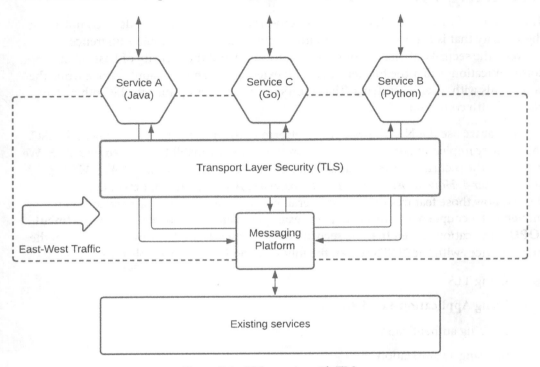

Figure 7.6 – ISC security with TLS

As shown in the preceding diagram, each microservice communicates with the messaging platform (NATS server) via TLS, and the NATS servers are enabled to communicate over TLS. We will look at how this can be done (with an example) in the next section when we talk about using NATS for securing ISC. This option is suitable if we have network-level isolation and security for microservices deployment. Usually, if microservices are deployed in such a secured network zone, we could consider both NATS and the microservices as trusted subsystems and enable TLS-only security for ISC. This would reduce the additional performance overhead that would otherwise be added to the request processing.

Security with authentication

Sometimes, securing ISC with TLS may not be enough due to the strict security measures applied by certain industries that are required to comply with various regulations. In such cases, NATS allows the implementation of additional security using authentication, in addition to the TLS-based communication between the microservices and the NATS server (messaging platform). The following diagram depicts this concept:

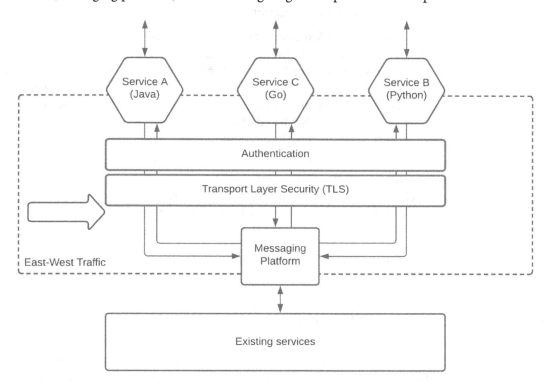

Figure 7.7 – Inter-service security with TLS and authentication

The preceding diagram depicts how microservices communicate with the messaging platform over a secured channel that has authentication implemented on top of the standard TLS-based security. NATS supports different options to implement authentication based on the requirements of the use case. It supports the following options:

- **Token authentication**: This is a simple mechanism to authenticate clients (microservices) to connect with the NATS server using a string (token) that is preconfigured in the NATS server. It does not recognize a specific user or application.

- **Basic authentication**: This method allows clients to be authenticated with a username and password pair that is configured in the server. It allows the server to identify the client (user) separately and provides additional control over the usage.

- **Certificate-based authentication**: The server can verify the client's TLS certificate and validate that against a known or trusted **Certificate Authority (CA)**. In addition to that, the NATS server can extract information from the certificate to validate the user for additional authentication.

- **Key-based authentication**: NATS supports an NKeys-based approach to authenticate a client. In this approach, the client needs to provide its public key to the server and then digitally sign a challenge that is offered by the server for each new connection with its private key. The server will validate the generated signature against the public key provided by the client. This is a better approach than token authentication, where keys need to be stored at the server side and the client needs to send that in plain text to connect with the server.

- **JWT-based authentication**: In this approach, users can be managed separately from the NATS server, and clients will acquire a JWT token from a server that is trusted by the NATS server. Once a client connects to the NATS server with the JWT token, the NATS server validates the client using a trust chain, which will eventually validate the server that provided the JWT token to the client.

We will discuss these methods with a few examples in the upcoming sections within this chapter.

Security with authorization

As we were able to observe in the previous section on authentication, some authentication mechanisms do not allow the client to be identified separately. This might cause problems in situations where different clients (microservices) need to be secured with distinct levels of security. This is typically called authorization, where the types of tasks an authenticated user can execute on a system are defined. The following diagram illustrates this concept:

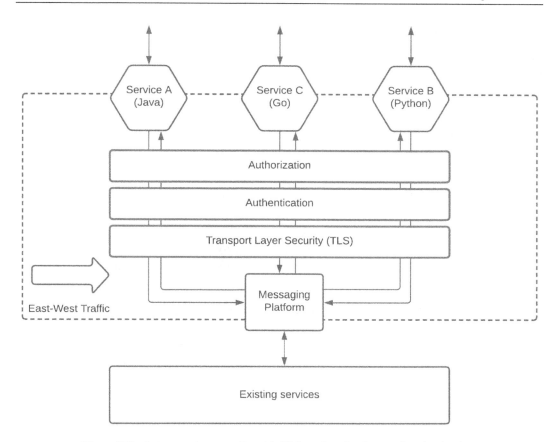

Figure 7.8 – Inter-service security with TLS, authentication, and authorization

As depicted in the preceding diagram, authorization adds an extra layer of security on top of the standard TLS and authentication-based security by restricting the users based on their identity instead of providing the same level of privileges to all the users (clients). In the context of microservices, different microservices might require diverse levels of permissions on the subjects. The NATS server supports authorization using subject-level permissions based on the specified users' list. These permissions can define the subjects that a user can publish to and/or subscribe to. Now that we have a good understanding of the different security levels supported by the NATS server for ISC and how it relates to a microservices architecture overall, we will take a look at some examples to understand how to secure microservices with NATS.

Using NATS to secure ISC

In this section, we will learn to secure microservices with the NATS server by configuring the NATS server and implementing code samples to try out different options discussed in the preceding sections.

Implementing TLS with NATS for microservices

Let's set up a NATS server cluster with TLS and connect to it from our sample code.

Setting up a NATS cluster with TLS

The TLS configuration for the NATS server consists of two separate sections to secure client-server communication as well as server-server communication for clustering. In addition to that, the NATS server monitoring port also needs to be configured for TLS-based security. The steps mentioned next allow you to configure NATS servers with TLS security:

1. Since we are setting up a test environment, we will use a self-signed certificate to implement TLS security. To create certificates, we will be using the **cfssl** tool maintained by Cloudflare (`https://github.com/cloudflare/cfssl`). You can install this tool via a binary or via the `go` command, as mentioned next:

    a) Installing via binaries (download the related binary), available at `https://github.com/cloudflare/cfssl/releases`

    b) Installing via the Go tool, as follows:

    ```
    $ go install github.com/cloudflare/cfssl/cmd/cfssl@latest
    ```

2. Once the cfssl tool is installed on the system, let's go ahead and create security roles for the certificates that we are going to generate. We will be generating certificates for the server and the route roles, as follows:

    a) **Server**: The certificate used for client-server TLS communication

    b) **Route**: The certificate used for server-server clustering TLS communication

3. Let's go ahead and create a directory called `certs` in whichever location you are executing these commands and change the directory to that folder. Inside that folder, create a `ca-config.json` file with the content shown next:

    ```
    {
      "signing": {
        "default": {
          "expiry": "10000h"
    ```

```
},
"profiles": {
"server": {
"expiry": "10000h",
"usages": [
"signing",
"key encipherment",
"server auth"
]
},
"route": {
"expiry": "10000h",
"usages": [
"signing",
"key encipherment",
"server auth",
"client auth"
]
}
}
}
}
```

4. The next step in setting up the self-signed certificates is to create a custom root CA. To do that, you can create another file inside the certs directory with the name ca-csr.json and add the following content to it:

```
{
"CN": "NATS Book CA",
"key": {
"algo": "rsa",
"size": 2048
},
"names": [
{
"C": "US",
"L": "CA",
```

```
"O": "NATS Book Company",
"ST": "San Francisco",
"OU": "Org Unit 1"
}
]
}
```

5.   Now, with these two files in our `certs` directory and the cfssl tool installed in the system, we can generate a CA certificate using the following command:

```
$ cfssl gencert -initca ca-csr.json | cfssljson -bare ca
```

This command will generate three files in the same directory, named `ca.pem`, `ca-key.pem`, and `ca.csr`.

6.   The next step is to generate server certificates that are going to be used to secure client-server communication. You can create a new file called `server.json` and add the following content to it:

```
{
"CN": "nats server",
"hosts": [
"*.nats.example.com",
"nats.example.com"
],
"key": {
"algo": "rsa",
"size": 2048
},
"names": [
{
"C": "US",
"L": "CA",
"ST": "San Francisco"
}
]
}
```

7. Execute the following command to generate a server certificate using the `server.json` file:

```
$ cfssl gencert -ca=ca.pem -ca-key=ca-key.pem -config=ca-config.json -profile=server server.json | cfssljson -bare server
```

Once the command is executed, it will generate three more files in the same directory, named `server.pem`, `server-key.pem`, and `server.csr`. We now have the server certificates to secure client-server communication.

8. The next step is to generate certificates to secure server-server cluster communication. For this, we will be generating a wildcard certificate with a subdomain (`nats-route`). Go ahead and create another file in the same `certs` directory with the name `routes.json`, and add the following content:

```
{
"CN": "nats route",
"hosts": [
"*.nats-route.example.com"
],
"key": {
"algo": "rsa",
"size": 2048
},
"names": [
{
"C": "US",
"L": "CA",
"ST": "San Francisco"
}
]
}
```

9. To create cluster certificates, you can execute the following command:

```
$ cfssl gencert -ca=ca.pem -ca-key=ca-key.pem -config=ca-config.json -profile=route routes.json | cfssljson -bare route
```

This command will create three files named `route.pem`, `route-key.pem`, and `route.csr` in the same directory.

10. We now have all the certificates needed to secure the NATS server with TLS. Let's go ahead and configure the servers and start the three-node cluster. To do that, we can create three configuration files—`node1.conf`, `node2.conf`, and `node3.conf`—in the same `certs` directory to start the servers, as mentioned next.

This is the configuration file (`node1.conf`) for node 1:

```
listen: 127.0.0.1:4222

tls {
cert_file = './server.pem'
key_file = './server-key.pem'
ca_file = './ca.pem'
timeout = 5
}

cluster {
name: sample-cluster
listen: 127.0.0.1:4248
tls {
cert_file = './route.pem'
key_file = './route-key.pem'
ca_file = './ca.pem'
timeout = 5
}
routes = [
nats-route://node-B.nats-route.example.com:5248
nats-route://node-C.nats-route.example.com:6248
]
}
```

This is the configuration file (`node2.conf`) for node 2:

```
listen: 127.0.0.1:5222

tls {
cert_file = './server.pem'
```

```
key_file = './server-key.pem'
ca_file = './ca.pem'
timeout = 5
}

cluster {
name: sample-cluster
listen: 127.0.0.1:5248
tls {
cert_file = './route.pem'
key_file = './route-key.pem'
ca_file = './ca.pem'
timeout = 5
}
routes = [
nats-route://node-A.nats-route.example.com:4248
nats-route://node-C.nats-route.example.com:6248
]
}
```

This is the configuration file (node3.conf) for node 3:

```
listen: 127.0.0.1:6222

tls {
cert_file = './server.pem'
key_file = './server-key.pem'
ca_file = './ca.pem'
timeout = 5
}

cluster {
name: sample-cluster
listen: 127.0.0.1:6248
tls {
cert_file = './route.pem'
key_file = './route-key.pem'
ca_file = './ca.pem'
```

```
timeout = 5
}
routes = [
nats-route://node-A.nats-route.example.com:4248
nats-route://node-B.nats-route.example.com:5248
]
}
```

Let's look at these files in a bit more detail to understand what we are configuring here. We first specify the host and port to listen on client connections and then configure the TLS certificates for client-server communication by providing the certificate paths and the connection timeout. Then, in the `cluster` section, we define the cluster name, cluster host, port, and the TLS configurations for route connections that are used for server-server communication. Finally, in the `routes` section, we provide the connectivity details of the other servers by specifying the hostname and port that are exposed for cluster communication.

In an actual setup, we could use the same port for cluster communication instead of the different ports we have used here (for example, `4248`, `5248`, and `6248`). Since we are testing our cluster on the same computer, we need to use different ports to avoid port conflicts.

11. One last thing to do before starting the server is to configure the `/etc/hosts` file to identify the hostnames that we have used in the configuration files. You can add the following entries to the `/etc/hosts` file to configure it:

```
127.0.0.1        node-A.nats-route.example.com
127.0.0.1        node-B.nats-route.example.com
127.0.0.1        node-C.nats-route.example.com
127.0.0.1        nats.example.com
```

12. We now have all the necessary configurations to try out the NATS cluster setup with TLS enabled. Let's go ahead and start the three servers by executing the commands mentioned next. You should execute these commands from the `certs` directory that we created at the beginning of this section in three separate terminal (or command) windows:

a) For terminal 1, run the following command:

```
$ nats-server -config node1.conf -D
```

b) For terminal 2, run the following command:

```
$ nats-server -config node2.conf -D
```

c) For terminal 3, run the following command:

```
$ nats-server -config node3.conf -D
```

When the three servers are starting, you will see a few error messages printed in the logs at the beginning until all three servers are started, since each server is trying to connect with the other servers. These errors go away when all three servers are started.

13. We now have our NATS server cluster started with TLS enabled for client-server and server-server communication. Let's go ahead and send a message to this cluster using a sample Go program over a TLS connection. The following program (certs/subscriber/main.go) listing can be used to subscribe to a message on the patient.profile subject by connecting to the cluster:

```
package main
import (
"fmt"
"log"
"sync"
"github.com/nats-io/nats.go"
)

func main() {
// Create a connection to the NATS server over TLS with
the RootCA
nc, err := nats.Connect("tls://nats.example.com:4222
,tls://nats.example.com:5222,tls://nats.example.com:
6222", nats.RootCAs("../ca.pem"))
if err != nil {
log.Fatal(err)
}
defer nc.Close()

// Use a WaitGroup to wait for a message to arrive
wg := sync.WaitGroup{}
wg.Add(1)
```

```
fmt.Println("Listening on [patient.profile] subject")


// Subscribe
if _, err := nc.Subscribe("patient.profile", func(m
*nats.Msg) {
wg.Done()
fmt.Printf("Received on [%s]: '%s'", m.Subject, string(m.
Data))
}); err != nil {
log.Fatal(err)
}


// Wait for a message to come in
wg.Wait()
}
```

In the preceding code segment, we connected to the NATS server cluster using the TLS protocol and then subscribed to the messages published on the patient. profile subject.

14. You can run the preceding program by going into the subscriber directory within the certs directory and executing the following command (make sure you have the go.mod file in the same directory):

```
$ go run main.go
```

15. Then, we can publish a message on the same patient.profile subject by connecting with another client over TLS. The following code sample (certs/publisher/main.go) can be used to publish a message over a TLS connection:

```
package main
import (
"log"
"github.com/nats-io/nats.go"
)


func main() {
// Create a connection to the NATS server over TLS with
the RootCA
```

```go
nc, err := nats.Connect("tls://nats.example.
com:4222,tls://nats.example.com:5222,tls://nats.example.
com:6222", nats.RootCAs("../ca.pem"))
if err != nil {
log.Fatal(err)
}
defer nc.Close()


// Publish a message on "patient.profile" subject
subj, msg := "patient.profile", []
byte("{\"name\":\"parakum\"}")
nc.Publish(subj, msg)
nc.Flush()
if err := nc.LastError(); err != nil {
log.Fatal(err)
} else {
log.Printf("Published [%s] : '%s'\n", subj, msg)
}
}
```

In the preceding code sample, we are connecting to the NATS server cluster over TLS and then publishing a message on the patient.profile subject.

16. The preceding code can be executed by going inside the publisher directory within the certs directory and executing the following command:

```
$ go run main.go
```

Once you run this publisher client program, you should observe that the message is received by the subscriber program and the communication is successful. You can try out the same two samples by shutting down a couple of NATS servers and still observe that the communication is successful if there is at least one server available, regardless of which one it is.

We have now configured a NATS server cluster to communicate over TLS for both client-server and server-server communication. Let's take a brief look at implementing authentication and authorization on top of this configuration, with a few examples.

Implementing basic authentication and authorization with NATS

NATS allows users to implement authentication for client-server communication with different approaches. Let's discuss how we can implement basic authentication along with authorization for our OPD application, as follows:

1. The first thing we must do is configure the NATS servers with the required configurations to validate the users and their permissions. We can modify the server configuration files that we created in the previous section with the authorization details shown here:

```
authorization {
default_permissions = {
publish = "SANDBOX.*"
subscribe = ["PUBLIC.>", "_INBOX.>"]
}
ADMIN = {
publish = ">"
subscribe = ">"
}
REGISTRATION = {
publish = "patient.register"
}
INSPECTION = {
subscribe = "patient.register"
publish = "patient.treatment"
}
TREATMENT = {
subscribe = "patient.treatment"
publish = "patient.release"
}
RELEASE = {
subscribe = "patient.release"
}
users = [
{user: registration_service, password: $REGISTRATION_
PASS, permissions: $REGISTRATION}
```

```
{user: inspection_service, password: $INSPECTION_PASS,
permissions: $INSPECTION}
{user: treatment_service, password: $TREATMENT_PASS,
permissions: $TREATMENT}
{user: release_service, password: $RELEASE_PASS,
permissions: $RELEASE}
{user: admin, password: $ADMIN_PASS, permissions: $ADMIN}
{user: other, password: $OTHER_PASS}
]
}
```

Here, we define permissions for each microservice and then assign those permissions to the individual users that represent microservices. You can find the updated server configuration files in the `certs` directory, with the names `node1-auth.conf`, `node2-auth.conf`, and `node3-auth.conf`.

2. Before starting the servers, we need to configure the passwords for each user. You can set the passwords as environment variables, as shown in the following code snippet. You need to set these environment variables in all three terminals that are used to start the NATS servers:

```
export REGISTRATION_PASS=reg123;
export INSPECTION_PASS=ins123;
export TREATMENT_PASS=trt123;
export RELEASE_PASS=rel123;
export ADMIN_PASS=admin123;
export OTHER_PASS=other123;
```

3. Let's go ahead and start the NATS server cluster with these updated configuration files, as follows:

a) For terminal 1, run the following command:

```
$ nats-server -config node1-auth.conf -D
```

b) For terminal 2, run the following command:

```
$ nats-server -config node2-auth.conf -D
```

c) For terminal 3, run the following command:

```
$ nats-server -config node3-auth.conf -D
```

4. Let's go ahead and implement a sample publisher and subscriber to verify the authorizations settings in the server. The only difference between the publisher code we used in the previous section and this one is the way we connect to the NATS server cluster. The following code segment (`certs/publisher-auth/main.go`) can be used to connect to the NATS server cluster, which is configured with authorization:

```go
nc, err := nats.Connect("tls://nats.example.
com:4222,tls://nats.example.com:5222,tls://nats.example.
com:6222",
nats.RootCAs("../ca.pem"), nats.UserInfo("registration_
service", "reg123"))
if err != nil {
log.Fatal(err)
}
defer nc.Close()

// Publish a message on "patient.profile" subject
subj, msg := "patient.register", []
byte("{\"name\":\"parakum\"}")
nc.Publish(subj, msg)
nc.Flush()
if err := nc.LastError(); err != nil {
log.Fatal(err)
} else {
log.Printf("Published [%s] : '%s'\n", subj, msg)
}
```

5. Then, we can implement sample code (`cert/subscriber-auth/main.go`) to subscribe to this `patient.register` subject with the relevant authentication details, as shown here:

```go
nc, err := nats.Connect("tls://nats.example.
com:4222,tls://nats.example.com:5222,tls://nats.example.
com:6222",
nats.RootCAs("../ca.pem"),nats.UserInfo("inspection_
service", "ins123"))
if err != nil {
log.Fatal(err)
}
```

```
defer nc.Close()

// Use a WaitGroup to wait for a message to arrive
wg := sync.WaitGroup{}
wg.Add(1)
fmt.Println("Listening on [patient.register] subject")

// Subscribe
if _, err := nc.Subscribe("patient.register", func(m
*nats.Msg) {
wg.Done()
fmt.Printf("Received on [%s]: '%s'", m.Subject, string(m.
Data))
}); err != nil {
log.Fatal(err)
}

// Wait for a message to come in
wg.Wait()
```

6. We now have everything ready to test out the NATS cluster that is secured with TLS and user-based authorization. Let's go ahead and start our subscriber code by going into the /certs/subscriber-auth directory and executing the following command:

```
$ go run main.go
```

7. Then, we can start the publisher sample by going into the /certs/publisher-auth directory and executing the following command:

```
$ go run main.go
```

The sample will publish a message to the patient.register subject, which will eventually be received by the subscriber we started in the previous step. When these two examples are executed, the subscriber sample will print the following log in the console:

```
Received on [patient.register]: '{"name":"parakum"}'
```

8. The next step is to add these improvements to our OPD application. That is left as an exercise for you. The full source code for these examples can be found at `https://github.com/PacktPublishing/Designing-Microservices-Platforms-with-NATS/tree/main/chapter7`.

Now that we have implemented security for ISC using NATS with a few examples, the next step is to implement security for North-South traffic. We will discuss that in *Chapter 9, How Microservices and NATS Co-Exist with Integration Platforms?*. We will use the API gateway-based approach that we discussed at the beginning of this chapter to implement security for North-South traffic.

Summary

In this chapter, we discussed how security in microservices can be implemented for both North-South traffic and East-West traffic. We started the chapter by discussing the different approaches we can follow to implement security for the external consumers of microservices. Then, we moved into the topic of ISC and discussed how NATS can be used to implement security for internal communications. Later in the chapter, we configured a three-node NATS server cluster with TLS security for both client-server and server-server communication with self-signed certificates. In the end, we added authentication and authorization on top of TLS, configured the NATS server cluster with additional security, implemented a few Go programs to connect with the NATS cluster, and verified the functionality. You have learned the security aspects of microservices architectures by reading the concepts presented in this chapter as well as by trying out the sample configurations and source code samples. Now, you can work on building secured microservices with NATS as the messaging layer for ISC.

In the next chapter, we will discuss how to implement monitoring for microservices using the NATS server for ISC.

8
Observability with NATS in a Microservices Architecture

Observability is a characteristic of a platform that defines how well the internal states of a system can be inferred from the knowledge of its external outputs. A platform without proper observability is hard to recover and troubleshoot when things do not go as expected. In a world of distributed systems, failure is inevitable, and systems need to be designed in such a way that they can withstand failures. It is the responsibility of each microservices team to have proper observability in their implementations and share a common set of tools to monitor the applications.

Given that a microservices architecture encourages building more services, it becomes a challenge to build correlations among messages when communicating across multiple services. With the usage of NATS for interservice communication, it becomes important that the NATS server has proper mechanisms to monitor the message interactions. Since all the messages are transferring through the NATS server, having proper observability is crucial in recovering quickly and reducing business impact in failure scenarios.

In this chapter, we are going to discuss the following main topics:

- Observability in the microservices context
- Observability features of NATS
- How to use NATS observability in a microservices architecture

By the end of this chapter, you will understand the concept of observability in a microservices context and how to build an observable microservices architecture with the help of the NATS platform.

Technical requirements

In this chapter, we will be configuring the NATS server to enable observability of the interservice communication and improve our sample microservices to include observability features. The following software components need to be installed to try out the examples mentioned in this chapter:

- The Go programming language
- The NATS server
- Prometheus
- Grafana
- Loki and Promtail

The source code of the examples used in this chapter can be found at `https://github.com/PacktPublishing/Designing-Microservices-Platforms-with-NATS/tree/main/chapter8`.

Observability in a microservices context

In the first few chapters of this book, we discussed the advantages of a microservices architecture and how it allows applications to be developed as a set of independent components that work cohesively. However, we have overlooked one thing, which is the complexity it adds to the process of monitoring and recovery from failure. One thing that we want to make clear at the beginning of this chapter is that monitoring is not the same as observability. While observability is a feature of the overall system, monitoring is a process that we execute to capture the details (external outputs). Having said that, both observability and monitoring are interconnected.

In a monolithic application, it is not that difficult to identify the root causes of a failure since there is only one place to look. From that place, we can dig into the details and identify the root cause. In the real world, this may not be as easy as it sounds. But comparatively, it is much easier than with microservices where you have tens, hundreds, or even thousands of services that can cause the issue and you have no idea where to begin. In both scenarios, having proper observability in the service (monolith) or services (microservices) is critical in identifying the root cause of a failure.

Let's first identify what aspects we need to monitor in a microservices (or any distributed application) architecture that would help us to keep the system available and avoid any business impact, or to minimize downtime. Here is a list of things that we should monitor:

- **Server (host) monitoring**: This is about monitoring the status of the server so that it is performing at a good level without underutilization or overutilization. We can monitor the CPU usage, memory usage, and load average on the server against a certain threshold value. Typically, we consider a server as overutilized if the CPU usage goes over 60–70%.

- **Application (microservices) monitoring**: This is where we monitor the application-level statistics (metrics), such as latency, error rates, and request rates, so that we can make decisions on when to scale out the services based on the demand from the consumers.

- **Log monitoring**: This is where most of the observability-related information is monitored and accessed. Microservices will utilize logging to output the details of the system state via different log entries using different categories, such as INFO, ERROR, and DEBUG, which will then be used to infer the state of the system and identify the root causes of failures.

- **Health monitoring**: This is where the application's health is monitored continuously through heartbeats, alerts, and notifications. The heartbeat check is the monitoring endpoint used by front-facing components such as load balancers to verify the availability of the server or the service itself. In most cases, this is a separate endpoint per server and/or per service that responds immediately with a simple response so that the client (for example, a load balancer) can verify the status. These health checks are run periodically at the load balancer level so that it routes the requests only to the available services. In addition to the heartbeat check, there are mechanisms used by applications to monitor abnormal activities such as high CPU usage, high memory usage, and frequent error logs. They then generate alerts and notifications so that the operations teams can take action on those notifications to rectify the system and avoid future failures.

If we think about building a distributed application with a couple of microservices running on a couple of servers, monitoring the overall system may not be that hard. But in the real world, we need to design applications that span across tens or hundreds of servers. Microservices architecture is designed to support that level of scalability based on consumer demand.

In such a scenario, let's discuss how observability and monitoring can be implemented so that we have a better chance of recovering quickly from failures. The following figure depicts a typical microservices-based deployment with the monitoring aspects that we discussed in the previous section:

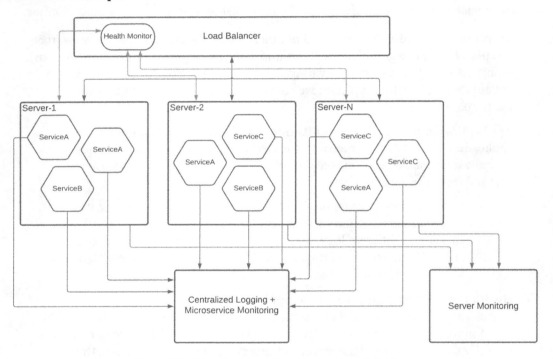

Figure 8.1 – Monitoring a microservices-based platform

The preceding figure depicts a use case where a microservices-based application is deployed in a typical **virtual machine** (**VM**) or physical hardware-based environment. In this scenario, multiple copies of the same microservice are deployed in different servers to fulfill the availability and scalability demands.

The load balancer is used to distribute the load among different instances of the microservices, and it contains a health monitor task that verifies the health of the servers as well as the health of the individual microservices by calling the health check APIs. Typically, this API is implemented by each microservice in a different context, such as / healthz, so that the load balancer can periodically check the status of the service.

The server monitoring needs to be done at a central place so that the operations teams can identify the status of the overall deployment and take actions on individual servers accordingly based on the overall demand of the consumers. Here the details are monitored per server, such as CPU, memory, and load average.

To reduce the complexity of *Figure 8.1*, we have kept the application or microservices monitoring and centralized logging components in the same box. These two components can live as separate components in real-world applications if required. Centralized logging allows us to manage the observability information in a central place and perform common operations that are typically required for troubleshooting.

Instead of using different tools for different microservices, all the teams can agree on a single logging technology and a standard logging format so that correlating the logs is easier when things start to fail at the application level. These logs need not be written to the same log files by different microservices. But all the microservices can use a common correlation ID when printing log entries and pass it through the message body or as a header to the next microservice, so that each microservice is aware of this common ID.

This correlation ID can come in a request or a header from the client. If it is not available, then each microservice can check the availability of the ID and generate it, and pass it to the subsequent services down the line so that they can reuse that ID. In addition to that, each microservice instance can have its own unique ID so that debugging and troubleshooting becomes much easier. The following figure depicts a scenario where the **CorrelationID** identifier sent by the client is used to enable observability for microservices:

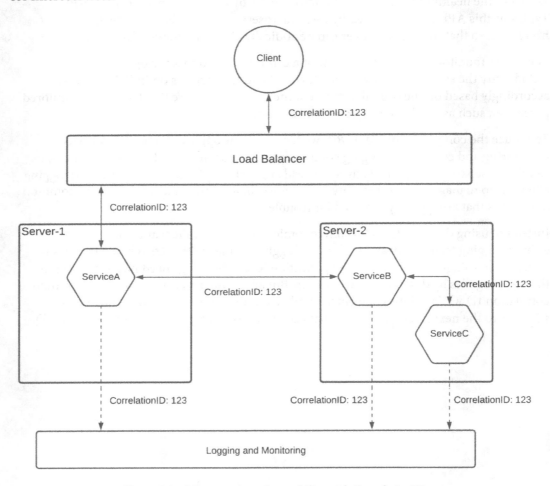

Figure 8.2 – Microservices observability with CorrelationID

As depicted in the preceding figure, the client sends a CorrelationID identifier with the value **123** to the load balancer and it passes that through to the microservices layer. In the microservices layer, **ServiceA** uses that value to print logs if required and pass it to the **ServiceB** where it does the same and passes it to **ServiceC**. If something goes wrong at any of the microservices, we can trace back the request with a CorrelationID identifier and find out the erroneous microservice, and then take actions to fix it. Given that we have a direct correlation with the client in this scenario, we can take actions such as message replay in this scenario.

If the client is not interested in sending a CorrelationID identifier, then ServiceA can generate the ID and pass that to ServiceB so that the remaining process that is depicted in *Figure 8.2* is executed. In this scenario, we won't be able to replay the messages since we don't have a direct mapping of client messages along with the internal CorrelationID identifier. But this is enough for most applications.

The usage of CorrelationID is a common practice that can be used in different deployment environments, including VMs, physical data centers, and containerized deployments. Containerized environments provide observability features that are built into the platform so that users can utilize those in addition to the features implemented at the application level. Given that microservices are often deployed in containerized environments, the following section provides a brief overview of observability in containerized deployments.

Observability in containerized deployments

Given the fact that microservices-based solutions are often deployed in containerized environments (for example, Docker) with container orchestration platforms such as Kubernetes, the monitoring of underlying server or VM infrastructure is abstracted away from the users. As an example, Kubernetes provides its own cluster monitoring and managing dashboards so that users can easily monitor the server-level details without using any additional tools.

In Kubernetes, each microservice runs on a separate Pod, and monitoring of the Pod-related activities is done through the monitoring tools provided by Kubernetes itself. It also provides the ability to configure server health checks with its liveness and readiness probes so that load balancers can verify the availability of the Pods that run microservices internally. But still, the application-level monitoring and log monitoring needs to be handled separately similar to what we discussed in the previous section with the usage of CorrelationIDs.

Best practices for microservices observability

Based on what we have discussed so far, we can identify a set of following best practices when implementing observability for microservices-based platforms:

- Instrument microservices code as much as possible with contextual details so that the failures can be traced back.

- It is always better to use CorrelationIDs when network interactions are happening in the code and put these details into a context.

- Always structure data that communicates over the wire and adheres to a common standard (for example, **Javascript Object Notation (JSON)**).

- Start instrumentation from the get-go and keep improving as the code becomes larger and complex. It is hard to introduce instrumentation at a later stage when the code becomes too large.

- Use log levels such as INFO, DEBUG, WARN, and ERROR to categorize log entries based on their severity.

- Use asynchronous logging mechanism so that the execution performance of the main program is not impacted by the logging.

- Emit log events with useful contextual information so that the person who reads these log entries can get value out of them.

So far, we have discussed the observability and monitoring aspects of microservices in general. In the upcoming sections of this chapter, we will discuss how NATS supports observability and helps us build the microservices architecture based on the NATS messaging platform.

Observability features of NATS

NATS provides two main methods to monitor the behavior of the NATS servers. Those are the following:

- Logging
- Monitoring

The internal implementation of the NATS server is done with a lot of instrumentation so that the users of the NATS server can monitor the servers in any deployment scenario. Let's discuss these options with a few examples.

Logging

Logging is the most common method of observability supported by the server software. It allows the users to track back the failures and possibly identify the root causes of such failures. There are many advanced log management tools available in the market so that users can monitor the logs in real time and trigger alerts and notifications based on certain log events, so that operations teams can take action on those alerts before the server goes into failure mode. The NATS server supports various logging options that you can enable through the command line or the configuration file.

When the NATS server starts with the default mode without any additional options in the command line or configuration file, it prints out the log entries to the console as well as to the default log file location, which is where you have installed the NATS server. You can change that by specifying the location of the log file when starting the NATS server through the command line or the configuration file. The following example shows how you can do that from the command line:

```
$ nats-server --log nats.log
```

Once you execute the preceding command, you no longer see the standard log entries that you observed in the terminal window before this. The reason for that is the log entries are now forwarded to the nats.log file, which we have specified. If you go ahead and look at the nats.log file with the following command, you should see the log entries related to server startup:

```
$ tail -f nats.log
[57001] 2021/08/04 16:55:23.282130 [INF]  Starting nats-server
[57001] 2021/08/04 16:55:23.282384 [INF]    Version:    2.3.0
[57001] 2021/08/04 16:55:23.282396 [INF]    Git:        [56a144a]
[57001] 2021/08/04 16:55:23.282404 [INF]    Name:
NC4ARPCQ2Y5GZ4OMC2LE3MLNSRR3TRZ6PC6VDSEH5PIY7F2JMJJ4S2LT
[57001] 2021/08/04 16:55:23.282412 [INF]    ID:
NC4ARPCQ2Y5GZ4OMC2LE3MLNSRR3TRZ6PC6VDSEH5PIY7F2JMJJ4S2LT
[57001] 2021/08/04 16:55:23.288322 [INF] Listening for client
connections on 0.0.0.0:4222
[57001] 2021/08/04 16:55:23.289009 [INF] Server is ready
```

Similar to this, there are other options available so that we can enable additional levels of logging, such as debug logs, trace logs, and syslog based on your needs. It also provides features such as log rotation based on the log file size, so that older files can be backed up to save space.

More information on configuring the logging for the NATS server can be found in the official NATS documentation available at `https://docs.nats.io/nats-server/configuration/logging`.

Monitoring

The NATS server follows a modern approach to monitoring the server in addition to the traditional log-based monitoring by running an HTTP server that exposes server details via a REST API. This API is exposed over HTTP by default, and we can expose it over HTTPS for additional security. We can configure the port of this HTTP server via the configuration file or the command line.

Let's go ahead and start the NATS server with HTTPS-based monitoring. To do that, we can reuse the same set of certificates that we had generated in *Chapter 7, Securing a Microservices Architecture with NATS* You can check out the code at the GitHub repository: `https://github.com/PacktPublishing/Designing-Microservices-Platforms-with-NATS`.

Then, go inside the `chapter8` directory and update the `node1-auth.conf` file with the `https_port` parameter as follows:

node1-auth.conf

```
listen: 127.0.0.1:4222
https_port: 8222
tls {
cert_file = './server.pem'
key_file = './server-key.pem'
ca_file = './ca.pem'
timeout = 5
}
#### more...
```

Here, we are specifying the TLS configuration and the HTTPS monitoring port as `8222` and the client connection port as `4222`. Make sure the certificates generated in the previous chapter are copied into this directory before starting the server. Also, you should export the environment variables related to users and passwords with the following command:

```
$ export REGISTRATION_PASS=reg123;
export INSPECTION_PASS=ins123;
export TREATMENT_PASS=trt123;
export RELEASE_PASS=rel123;
export ADMIN_PASS=admin123;
export OTHER_PASS=other123;
```

Then we can start the NATS server with the following command:

```
$ nats-server --config node1-auth.conf
```

Now, you should be able to access the URL `https://127.0.0.1:8222` to view the monitoring details exposed by the NATS server. When you access this URL with a browser, you will get a browser warning since we are using a self-signed certificate. We can ignore this warning and proceed to the monitoring page.

Let's go ahead and make similar changes to the `node2-auth.conf` and `node3-auth.conf` files by adding `https_port` values `8223` and `8224` respectively. After that, go ahead and start nodes 2 and 3 to form the cluster by executing the following commands:

```
$ nats-server --config node2-auth.conf
$ nats-server --config node3-auth.conf
```

We can access the monitoring APIs of these two servers via the following URLs:

- `https://127.0.0.1:8223`
- `https://127.0.0.1:8224`

Now we have a 3-node NATS server cluster enabled with TLS for client connections and route connections with monitoring. Let's go ahead and run our publisher and subscriber code to make it interesting so that we have enough data to monitor through the monitoring endpoints. You can go inside the `chapter8/subs-1` directory and run the subscriber with the following command:

```
$ go run main.go
```

This command will start the subscriber that listens to the `patient.register` subject and is connected to the NATS server with port `5222`. You can start the other two subscribers located inside the `subs-2` and `subs-3` directories, which will listen to the same subject and are connected to the other two nodes on ports `6222` and `4222` respectively.

Finally, let's go ahead and start the publisher by going inside the `chapter8/publisher` directory and executing the same command as before.

Now we have the 3-node cluster enabled with TLS and monitoring with three subscribers and one publisher connected. Let's go ahead and access the monitoring endpoint (`https://localhost:8222`) to find out the information it exposes about the NATS server cluster.

This endpoint provides the details of the server via the following contexts:

- General server information (`/varz`)
- Connections-related information (`/connz`)
- Routing information (`/routez`)
- Information on gateways (`/gatewayz`)
- Information on leaf nodes (`/leafz`)
- Subscription routing information (`/subsz`)
- Account information (`/accountz`)
- JetStream information (`/jsz`)

All the aforementioned endpoints provide a JSON response to the client and support JSONP and CORS, so that you can create monitoring web applications using a **single page application** (**SPA**) approach. Let's discuss some of the important aspects of these endpoints related to our topic.

General server information (/varz)

This endpoint provides general information on the server and the status, which includes details such as the following:

- Server name, ID, TLS, and authorization details
- Cluster details and routes
- CPU and memory usage data
- Connection URLs
- Hostname, ports, and timeouts

- In/out messages and total size of messages
- Subscriptions

If you access the `https://localhost:8222/varz` URL in your setup, you should get a response similar to the following content:

```
{
    "server_id": "NAN5TOLVVYVTAFZBTPMZ7NIXUF5SQKPFFWODFY327RRHBW3DJCRGLF3J",
    "server_name": "NAN5TOLVVYVTAFZBTPMZ7NIXUF5SQKPFFWODFY327RRHBW3DJCRGLF3J",
    "version": "2.3.0",
    "proto": 1,
    "git_commit": "56a144a",
    "go": "go1.16.5",
    "host": "127.0.0.1",
    "port": 4222,
    "auth_required": true,
    "tls_required": true,
    "connect_urls": [
      "127.0.0.1:4222",
      "127.0.0.1:5222",
      "127.0.0.1:6222"
    ],
    "max_connections": 65536,
    "ping_interval": 120000000000,
    "ping_max": 2,
    "http_host": "127.0.0.1",
    "http_port": 0,
    "http_base_path": "",
    "https_port": 8222,
    "auth_timeout": 6,
    "max_control_line": 4096,
    "max_payload": 1048576,
    "max_pending": 67108864,
    "cluster": {
      "name": "sample-cluster",
      "addr": "127.0.0.1",
      "cluster_port": 4248,
      "auth_timeout": 6,
      "urls": [
        "node-B.nats-route.example.com:5248",
        "node-C.nats-route.example.com:6248"
      ],
      "tls_timeout": 5,
      "tls_required": true,
      "tls_verify": true
    },
    "gateway": {},
    "leaf": {},
    "jetstream": {},
    "tls_timeout": 5,
    "write_deadline": 10000000000,
    "start": "2021-09-21T05:29:22.04456Z",
    "now": "2021-09-23T04:26:07.921918Z",
    "uptime": "1d22h56m45s",
    "mem": 16248832,
    "cores": 12,
    "gomaxprocs": 12,
    "cpu": 0,
    "connections": 2,
    "total_connections": 3,
    "routes": 2,
    "remotes": 2,
    "leafnodes": 0,
    "in_msgs": 4496,
    "out_msgs": 4496,
    "in_bytes": 2985022,
    "out_bytes": 2980160,
    "slow_consumers": 0,
    "subscriptions": 107,
    "http_req_stats": {
      "/connz": 3,
      "/routez": 1,
      "/varz": 2
    },
    "config_load_time": "2021-09-21T05:29:22.04456Z",
    "system_account": "$SYS"
}
```

Figure 8.3 – Output of the /varz endpoint in the NATS server

You should be able to identify the configurations related to this server instance with this endpoint.

Connections-related information (/connz)

Another important aspect of the NATS server that is exposed through the monitoring API is the connections-related information, such as current and closed connections with clients. This endpoint allows us to filter and sort the results based on various aspects. By default, it provides details of a maximum of 1,024 connections in a paging manner, and you can request more details by specifying the offset value from that page. The response from this endpoint provides information such as the following:

- Server ID
- Total connections on the server
- Details of each connection (in the response):

  - Connection ID/name
  - Connection port
  - In/out messages and sizes (total)
  - TLS details
  - Uptime/idle time/last activity time/start time

If you access the `https://localhost:8222/connz` URL in your setup, you should get a response similar to the following content:

```
{
  "server_id": "NAN5TOLVVYVTAFZBTPMZ7NIXUF5SQKPFFWODFY327RRHBW3DJCRGLF3J",
  "now": "2021-09-23T04:33:11.148249Z",
  "num_connections": 2,
  "total": 2,
  "offset": 0,
  "limit": 1024,
  "connections": [
    {
      "cid": 8,
      "ip": "127.0.0.1",
      "port": 56144,
      "start": "2021-09-21T05:39:51.991988Z",
      "last_activity": "2021-09-21T05:39:57.799091Z",
      "rtt": "204µs",
      "uptime": "1d22h53m19s",
      "idle": "1d22h53m13s",
      "pending_bytes": 0,
      "in_msgs": 0,
      "out_msgs": 1,
      "in_bytes": 0,
      "out_bytes": 18,
      "subscriptions": 1,
      "name": "inspection-service-3",
      "lang": "go",
      "version": "1.11.0",
      "tls_version": "1.3",
      "tls_cipher_suite": "TLS_AES_128_GCM_SHA256"
    },
    {
      "cid": 9,
      "ip": "127.0.0.1",
      "port": 56149,
      "start": "2021-09-21T05:39:57.795514Z",
      "last_activity": "2021-09-21T05:39:57.799091Z",
      "rtt": "163µs",
      "uptime": "1d22h53m13s",
      "idle": "1d22h53m13s",
      "pending_bytes": 0,
      "in_msgs": 1,
      "out_msgs": 0,
      "in_bytes": 18,
      "out_bytes": 0,
      "subscriptions": 0,
      "name": "registration-service",
      "lang": "go",
      "version": "1.11.0",
      "tls_version": "1.3",
      "tls_cipher_suite": "TLS_AES_128_GCM_SHA256"
    }
  ]
}
```

Figure 8.4 – Output of the /connz endpoint in the NATS server

Routing-related information (/routez)

This endpoint provides information on the cluster routes and their status. In our setup, we have three nodes clustered together, and each server has route connections from two other nodes. This response contains details such as the following:

- Server ID
- Number of routes
- Details of each route:
  - Route ID
  - Remote server ID
  - IP/port details
  - In/out messages and sizes of messages (total)

By accessing the `https://localhost:8222/routez` URL, we can see the following response from the monitoring endpoint related to the routes:

```
{
  "server_id": "NAN5TOLVVYVTAFZBTPMZ7NIXUF5SQKPFFWODFY327RRHBW3DJCRGLF3J",
  "now": "2021-09-21T05:41:49.197975Z",
  "num_routes": 2,
  "routes": [
    {
      "rid": 3,
      "remote_id": "NDSZ7JPO6BJRWRQRVH22A3KABPEWGY6P3GU3AXGARP4INOV6R6FUYWDX",
      "did_solicit": true,
      "is_configured": true,
      "ip": "127.0.0.1",
      "port": 55779,
      "pending_size": 0,
      "rtt": "149µs",
      "in_msgs": 58,
      "out_msgs": 58,
      "in_bytes": 37742,
      "out_bytes": 36379,
      "subscriptions": 35
    },
    {
      "rid": 5,
      "remote_id": "NAP7JSSD36Z4IOSVTDTK7YQW4AONIKHH3SSQX37TIMYZJMFFTVZQ5GUI",
      "did_solicit": true,
      "is_configured": true,
      "ip": "127.0.0.1",
      "port": 55799,
      "pending_size": 0,
      "rtt": "300µs",
      "in_msgs": 55,
      "out_msgs": 55,
      "in_bytes": 36469,
      "out_bytes": 35229,
      "subscriptions": 35
    }
  ]
}
```

Figure 8.5 – Output of the /routez endpoint in the NATS server

More details on these endpoints can be found on the official NATS documentation page at
`https://docs.nats.io/nats-server/configuration/monitoring`.

Now, we have a better understanding of the NATS observability features and how
microservices observability needs to be implemented. Let's discuss how we can use these
concepts to build a microservices-based platform with NATS using these concepts, by
using our sample OPD application and the NATS cluster.

Using NATS observability in a microservices architecture

So far, we have discussed microservice observability and NATS observability as two
separate topics. Let's aggregate these topics and produce a common approach to
implement observability for a microservices-based platform. The best way to learn
a concept is to try with a practical example. Let's build a simple application that has two
microservices written in Go. These microservices are called **publisher** and **subscriber**.
We are going to use the open source observability tools Prometheus, Grafana, and Loki in
this example.

The following figure shows the example application that we are going to build:

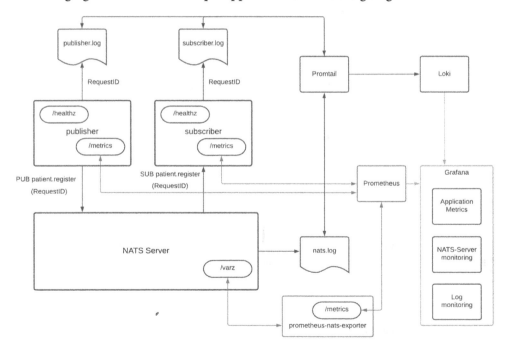

Figure 8.6 – Microservices observability example with Prometheus, Grafana, and Loki

The preceding figure depicts a simple application, consisting of a publisher microservice and a subscriber microservice interacting with each other using the NATS server. We have set up our monitoring ecosystem for this application using the following components:

- Prometheus
- Grafana
- Loki and Promtail

Let's get a basic understanding of these tools in the following subsections.

Prometheus

This is an open source monitoring solution that helps collect metrics on applications and provides interfaces to analyze, visualize, generate alerts, and take actions based on these alerts. Some key features include the following:

- A time series-based data collection
- The ability to run powerful queries to generate graphs, tables, and alerts
- Integration with visualization tools such as Grafana to generate powerful graphics
- Support for many programming languages with client libraries
- Simple implementation
- An excellent alerting capability with flexible queries

> **Important Note**
> We are using Prometheus to display observability aspects due to its popularity and wider adoption in the industry.

Grafana

This is an open source tool used to query, visualize, alert on, and understand observability data of applications regardless of the storage technology. It integrates with many metrics collecting systems, such as Prometheus, Loki, and Promtail. We can create, explore, and share observability data through nice dashboards, which are highly customizable. Some key features of Grafana include the following:

- The ability to unify data from different sources and storage technologies
- Flexibility and easy-to-use dashboard creation to share with other people

- Allowing data to be managed across teams and within teams with proper access control mechanisms

> **Important Note**
> We are using Grafana in this section due to its flexibility and wider adoption in the technology world.

Loki and Promtail

Loki is a log aggregation tool that collects logs from different applications for further analysis by effectively storing the log data. It uses indexing to improve the log analysis processes and uses labels to index the data without modifying the original log messages. It uses an agent such as Promtail to acquire the logs and turns the logs into streams, and then pushes them to Loki through an HTTP API. Some key features of Loki include the following:

- Effective log indexing with improved memory usage
- The ability to support large-scale applications with scalability
- Good integration with Grafana to build a complete observability stack
- Multi-tenancy to manage log data across teams

We are using Loki and Promtail in this section since it works well with Prometheus and Grafana.

Let's now go ahead and build our sample application. We are going to modify the publisher and subscriber samples we built in *Chapter 7, Securing a Microservices Architecture with NATS*. The full source code can be found at `https://github.com/PacktPublishing/Designing-Microservices-Platforms-with-NATS/tree/main/chapter8`.

Adding the health check API

The first thing we are going to do is implement the health check API for both publisher and subscriber examples. To do that, we are going to use the same `mux` library we used in our OPD application:

observability-sample/publisher/main.go

```
// ….. More code
// ListenAndServe takes the network address and port that
```

```go
// the HTTP server should bind to and starts it.
func ListenAndServe(addr string) error {
r := mux.NewRouter()
router := r.PathPrefix("/publisher/").Subrouter()

// Handle health check requests
router.HandleFunc("/healthz", HandleHealthCheck)

l, err := net.Listen("tcp", addr)
if err != nil {
return err
}
srv := &http.Server{
    Addr: addr,
    Handler: router,
    ReadTimeout: 10 * time.Second,
    WriteTimeout: 10 * time.Second,
    MaxHeaderBytes: 1 << 20,
}
go srv.Serve(l)
return nil
}

func HandleHealthCheck(w http.ResponseWriter, r
  *http.Request) {
    fmt.Fprintf(w, fmt.Sprintf("Service available. \n"))
}
// ….. More code
```

Here, we are starting an HTTP server with the /healthz endpoint to respond to health check calls with a simple response with HTTP status code 200. The same will be implemented for the subscriber sample as well.

Adding CorrelationID to the requests

The next thing we are going to do is generate a **Universal Unique Identifier (UUID)** and attach that to the request in the message body before publishing it to the NATS server. To do this, we are using a structure to define the message structure and including a field called `RequestID` to store the unique ID that we can use to trace the message in case of a failure:

observability-sample/publisher/main.go

```go
// .... More Code
// RegistrationRequest contains data about the patient.
type RegistrationRequest struct {
// Full Name of the patient.
FullName string `json:"full_name,omitempty"`
// Address of the patient.
Address string `json:"address,omitempty"`
// National Identification Number of the patient.
ID int `json:"id"`
// RequestID is the ID from the request.
RequestID string `json:"request_id,omitempty"`
}
// Create event
regEvent := RegistrationRequest{"Chanaka Fernando", "44
  Seeduwa", 1111, nuid.Next()}
reg_event, err := json.Marshal(regEvent)
if err != nil {
        log.Fatal(err)
}
// Publish a message on "patient.profile" subject
subj, msg := "patient.register", reg_event
nc.Publish(subj, msg)
// .... More Code
```

Here, we are creating the message with the auto-generated UUID and publishing that to the subject in the NATS server.

Adding improved logging for microservices

Each microservice should have its own log files to store the observability data. In addition to that, having separate log categories based on the details that are printed in the log makes analyzing and troubleshooting the issue scenarios much easier. Let's see how we can implement logging in our microservices:

observability-sample/publisher/main.go

```go
// ... more code
func initLoggers() {
  file, err := os.OpenFile("publisher.log",
    os.O_APPEND|os.O_CREATE|os.O_WRONLY, 0666)
  if err != nil {
     log.Fatal(err)
}
// more code
}
func main() {
// Initialize loggers
  initLoggers()
// more code
// Create event
  regEvent := RegistrationRequest{"Chanaka Fernando", "44
    Seeduwa", 1111, nuid.Next()}
  reg_event, err := json.Marshal(regEvent)
  if err != nil {
         ErrorLogger.Println(err)
         log.Fatal(err)
}
InfoLogger.Printf("Publishing message with ID %s",
  regEvent.RequestID)
}
// ... more code
```

In the preceding code segment, we are defining the log file name and the log categories with the format of the log entries. Then, we use these categories when logging the details of the program.

Collecting metrics and exposing them for application monitoring

Another important aspect of observability is application-level monitoring and metrics. We can use Prometheus's client exporter, which is a client library, to collect application metrics and expose them through a separate endpoint. Then, Prometheus can collect this information and provide advanced query capabilities and later publish it to `grafana` for better visualization. The following code sample shows how we can do that using our publisher example:

observability-sample/publisher/main.go

```go
// .. more code
var totalRequests = prometheus.NewCounterVec(
    prometheus.CounterOpts{
        Name: "http_requests_total",
        Help: "Number of get requests.",
    },
    []string{"path"},
)
// more code
func main() {
// Initialize Tracing
initTracing()
}
func prometheusMiddleware(next http.Handler) http.Handler {
return http.HandlerFunc(func(w http.ResponseWriter, r
  *http.Request) {
        route := mux.CurrentRoute(r)
        path, _ := route.GetPathTemplate()
        timer := prometheus.NewTimer(httpDuration.
          WithLabelValues(path))
        rw := NewResponseWriter(w)
        next.ServeHTTP(rw, r)
        statusCode := rw.statusCode
        responseStatus.WithLabelValues
          (strconv.Itoa(statusCode)).Inc()
          totalRequests.WithLabelValues(path).Inc()
```

```
         timer.ObserveDuration()
    })
}
```

In the preceding code segment, we are defining our own metrics to collect information related to totalRequests, responseStatus, and httpDuration so that we can measure the performance of our publisher application specific to HTTP calls. We intercept all the HTTP requests coming into the /publisher context with the middleware component, which collects this information and publishes it via the /metrics endpoint:

```
func NewResponseWriter(w http.ResponseWriter)
  *responseWriter {
    return &responseWriter{w, http.StatusOK}
}
func (rw *responseWriter) WriteHeader(code int) {
    rw.statusCode = code
    rw.ResponseWriter.WriteHeader(code)
}
func initTracing() {
    prometheus.Register(totalRequests)
    prometheus.Register(responseStatus)
    prometheus.Register(httpDuration)
}
// ListenAndServe takes the network address and port that
// the HTTP server should bind to and starts it.
func ListenAndServe(addr string) error {
    r := mux.NewRouter()
    router := r.PathPrefix("/publisher/").Subrouter()
    router.Use(prometheusMiddleware)
    router.Path("/metrics").Handler(promhttp.Handler())
// .. more code
```

Finally, we implement the response writing code and register the required details with Prometheus and start the microservice with the middleware component and the /metrics endpoint.

Exporting NATS server metrics

The next step is to expose the NATS server statistics to Prometheus using the monitoring endpoints provided. To achieve this, we are going to use the `promethus-nats-exporter` tool, which you can find at `https://github.com/nats-io/prometheus-nats-exporter`.

You can clone this repository to your working environment and then build it and install it, as mentioned in the `README` file. Once you have installed it, you can start the tool by running the following command:

```
$ prometheus-nats-exporter -varz http://localhost:8222
```

Here, we are specifying the monitoring port of the NATS server, which has already started. If you have not started the NATS server yet, you can start that with the following command:

```
$ nats-server --config node.conf --log nats.log
```

Make sure you copy the TLS configuration files you created in the `chapter7` folder to the `observability-sample` directory before running the NATS server.

Now, we are done with the implementation of the observability and publishing it through the metrics endpoints. Let's go ahead and configure the monitoring tools to consume this data and produce better analytics and visualization capabilities.

Starting microservices

Now, we have the observability features implemented in our publisher and subscriber applications. The next step is to go ahead and start the two applications by going inside the respective directories:

- **Terminal 1**:

```
$ cd chapter8/observability-sample/subscriber
$ go run main.go
```

- **Terminal 2**:

```
$ cd chapter8/observability-sample/publisher
$ go run main.go
```

Now, you should be able to observe the log files are populated with log entries. If you access the `/metrics` endpoints, you should be able to view the metrics for each microservice.

Configuring Prometheus

You can download and install Prometheus as per your operating system by visiting the website at `https://prometheus.io/download/`.

Once you install it, you can start Prometheus by specifying a configuration file to listen to certain applications. In our example, we are going to listen to our publisher, subscriber, and NATS server endpoints from Prometheus. The following listing can be used to configure Prometheus for that purpose:

observability-sample/prom1.yml

```yaml
global:
    scrape_interval: 15s
    evaluation_interval: 15s


scrape_configs:
    - job_name: prometheus
      static_configs:
        - targets: ['localhost:9090']
    - job_name: publisher
      metrics_path: /publisher/metrics
      static_configs:
        - targets:
          - localhost:9001
    - job_name: subscriber
      metrics_path: /subscriber/metrics
      static_configs:
        - targets:
          - localhost:9000
    - job_name: 'nats-test-server'
      static_configs:
        - targets: ['localhost:7777']
```

In the preceding configuration, we are specifying the applications that we need to observe along with the respective metrics endpoints. You can observe that we are configuring the NATS server with `localhost:7777`, which is the port exposed by the `prometheus-nats-exporter` tool.

Now you can go ahead and start Prometheus by going inside the installation directory with the following command:

```
$./prometheus --config.file prom1.yml
```

This will start the Prometheus server, and you can access it with `http://localhost:9090/`.

You can search certain metrics entries from this interface and visualize them with the Prometheus interface.

Configuring Loki and Promtail

The next step is to configure the log aggregation tool, which can help immensely in troubleshooting scenarios. You can download and install Loki and Promtail by following the instructions listed at `https://grafana.com/docs/loki/latest/installation/local/`.

Once you are done with the installation, you can start Loki by going inside the installation directory and executing the following command:

```
$./loki-darwin-amd64 -config.file=loki-local-config.yaml
```

Next, we need to configure Promtail to listen to the log files that are created by the applications and NATS server so that it can publish those log files as streams to Loki. The configuration of the Promtail local file is as follows:

promtail-local-config.yaml

```
server:
    http_listen_port: 9080
    grpc_listen_port: 0
positions:
    filename: /tmp/positions.yaml
clients:
    - url: http://localhost:3100/loki/api/v1/push

scrape_configs:
    - job_name: system
    static_configs:
    - targets:
```

```
  - localhost
  labels:
    job: varlogs
    __path__ : /var/log/*log
```

The preceding segment of the file is the default configuration to collect system logs and publish them to Loki. The following segments are used to collect logs from the publisher, subscriber, and NATS server:

```
- job_name: publisher
static_configs:
- targets:
  - localhost
  labels:
    job: publisher
    __path__ : /Users/chanakaf/Documents/source-code
      /Designing-Microservices-Platforms-with-NATS/
        chapter8/observability-sample/publisher
          /publisher.log
- job_name: subscriber
  static_configs:
  - targets:
    - localhost
    labels:
      job: subscriber
      __path__ : /Users/chanakaf/Documents/source-code
        /Designing-Microservices-Platforms-with-NATS/
          chapter8/observability-sample/subscriber
            /subscriber.log
- job_name: NATS-4222
  static_configs:
  - targets:
  - localhost
  labels:
    job: nats-4222
      __path__ : /Users/chanakaf/Documents/source-code
```

```
/Designing-Microservices-Platforms-with-NATS/
chapter8/observability-sample/nats.log
```

In the preceding sample configuration file, we have specified the log file location of the publisher, subscriber, and NATS server with the respective names so that we can search based on the application.

Configuring Grafana

The last component of our observability setup is Grafana, which is the data visualization tool. You can download and install Grafana as mentioned here: `https://grafana.com/docs/grafana/latest/installation/`.

Once it is started, you can access the interface by visiting `http://localhost:3000/`.

You can log in with the default username and password as `admin`. Next, you need to configure two separate data sources for Prometheus and Loki to monitor the metrics and logging from the same interface. You can follow the instructions given at the following link to add data sources: `https://grafana.com/docs/grafana/latest/datasources/add-a-data-source/`.

Now, we have everything set up to monitor and observe our microservices-based solution with NATS. You can import the `grafana-nats-dash.json` file to the Grafana server to create a graph to monitor the NATS server-related information. This file is available in the GitHub repository at `https://github.com/PacktPublishing/Designing-Microservices-Platforms-with-NATS/tree/main/chapter8/observability-sample`. You can play with the Grafana dashboards to create useful visualizations of the metrics and logging data. This part is kept as an exercise for you.

Summary

In this chapter, we discussed observability in the context of a microservices architecture and defined the distinct types of observability requirements in such a platform. We discussed how monitoring is related to observability and what types of information we can monitor in a microservices environment. Then, we discussed the various observability features available in the NATS platform and how that information can be accessed through log files and REST APIs. Finally, we built a comprehensive microservices-based observability platform with a sample application that uses NATS for interservice communication. We discussed setting up open source monitoring tools such as Prometheus, Grafana, and Loki to set up the observability platform for our sample application.

This chapter concludes our second section of the book, where we discussed the practical usage of NATS with microservices platforms by going through numerous examples and configurations. The next chapter starts our last section of the book, where we will discuss how microservices platforms along with NATS co-exist in enterprise IT environments and what the future of microservices and NATS looks like.

Section 3:
Best Practices
and Future
Developments

In this last section, we discuss how microservices-based applications integrate with existing systems and applications within an enterprise platform. We discuss some best practices to implement microservices in enterprise platforms with a few examples. We also discuss the latest trends in microservices architecture and how these new developments can help us to improve the proposed architecture. We also look at the new features of NATS messaging and how these features help us build an improved solution for future use cases.

This section contains the following chapters:

- *Chapter 9, How Microservices and NATS Co-Exist with Integration Platforms*
- *Chapter 10, Future of Microservices Architecture and NATS*

9
How Microservices and NATS Coexist with Integration Platforms

This chapter marks the beginning of the third and last section of this book, where we will discuss the best practices and future direction of microservices and NATS. We have discussed how microservices are built, secured, and observed along with NATS in the previous chapters of this book. In a typical **enterprise software system**, there are many other components that we need to consider when building microservice-based applications. Modern enterprise platforms are going through frequent changes and new systems need to coexist with older (legacy) systems at any given time. This is called the **brownfield** enterprise. A system built from scratch with microservices is called a **greenfield** enterprise, which is hard to find in real-world enterprise information systems.

The microservice platform that we design with NATS should be able to complement the overall enterprise platform, which consists of these various other systems. Integration platforms are common components that act as the core of the enterprise ecosystem. In this chapter, we will discuss how to build a cohesive system using microservices, NATS, and integration platforms.

In this chapter, we are going to discuss the following main topics:

- Understanding the brownfield enterprise software system

- How integration platforms work in an enterprise software system

- How microservices and NATS integrate with integration platforms

- How to build an enterprise platform with microservices, NATS, and an integration platform

By the end of this chapter, you will understand how enterprise software systems are built with integration platforms, microservices, NATS, and other software components in an effective way.

Technical requirements

In this chapter, we will be configuring the NATS server to work alongside microservices as well as an integration platform so that we can demonstrate the concepts provided. The following software components need to be installed for you to try out the examples mentioned in this chapter:

- The Go programming language

- The NATS server

- WSO2 API Manager

The source code for the examples in this chapter can be found at `https://github.com/PacktPublishing/Designing-Microservices-Platforms-with-NATS/tree/main/chapter9`.

Understanding the brownfield enterprise software system

With the popularity of the internet and the increased usage of consumer devices to access information, more and more businesses need to build enterprise software systems. These systems consist of various software applications that help in improving the overall efficiency of the business operations, all while providing great experiences to the customers. Different enterprises might require domain-specific software in addition to the common software systems that are used to process information and provide services to the customers. Let's take a look at some of the software systems that are used in most enterprise platforms:

- **Enterprise Resource Planning (ERP) system**: This is a system that's used to manage the enterprise resources and processes that are used for production, distribution, and other business operations.

- **Customer Relationship Management (CRM) system**: This is a system that's used to manage customer information and interactions with customers.

- **Databases**: These are used to store various types of data by applications that are built in-house, as well as other software components built by third parties.

- **Commercial Off-The-Shelf (COTS) applications**: These are applications that have been built specifically for a particular business domain and are available as installable software within the premises of the enterprise platform.

- **In-house built applications and tools**: These are software components that have been built by the internal IT teams for various requirements that cannot be fulfilled with the existing applications and systems.

- **Cloud services**: These are the services that are hosted and operated by the respective vendors in the cloud and purchased by the enterprises for requirements such as sales management, human resource management, expense management, and much more.

These different systems and applications cannot operate in isolation and must be integrated to build a fully functional, business-specific enterprise platform. In the beginning, we can use some of the capabilities built into these systems to integrate in a point-to-point manner. But that becomes complex and unmanageable with the increased number of systems and the variety of protocols and standards that are used in these applications to store and share information. To support these requirements, most enterprise software systems use components such as integration platforms and messaging platforms. These act as the intermediate components that orchestrate and mediate interactions among different systems to form a brownfield enterprise.

Now that we are aware of what a brownfield enterprise is, let's discuss how an integration platform can integrate the different components of a brownfield enterprise system.

How do integration platforms work in an enterprise software system?

The need for an integration platform arises from the business itself, as well as from the customers. From the customers' perspective, they would like to browse through all the possible prices, colors, sizes, and many other characteristics of the products and services a business offers. This requires extracting information from multiple systems and presenting that to the customer so that it looks like it came from just one system. Various business processes also need data to be shared and synchronized across different systems to make business decisions and measure the progress of the business. To fulfill all these needs, we require an integration platform. A typical integration platform provides these capabilities, such as the following:

- **Application integration**: Connecting different types of software applications to share and process data.

- **Data integration**: Integrate different data sources such as databases, files, and storage systems with each other for data synchronization and processing.

- **Business-to-Business (B2B) integration**: Connect with various systems and applications running on third-party partners and vendors to share business data.

- **API management**: Used to expose the applications and services in a secured and standard manner to external and internal users via different channels such as mobile, web, and terminal applications.

The aforementioned high-level capabilities of an integration platform can be further explained as follows:

- **Data transformation**: Convert data across different types and formats according to the applications and systems so that different systems can share and process data without any issue.

- **Protocol translation**: Different applications may use different protocols such as **HyperText Transfer Protocol (HTTP)**, **Advanced Message Queueing Protocol (AMQP)**, **Java Messaging Service (JMS)**, and **Health Level Seven (HL7)** to communicate with other applications, and not every application supports all the possible protocols. In such situations, the integration platform translates the protocols across formats so that systems can communicate with each other.

- **Application connectors**: There are different applications on board an enterprise platform as a business grows, and connecting to these various applications is a requirement for an integration platform so that new application onboarding does not take ages to install and configure to fit into the overall platform.

- **Routing and orchestration**: Another useful feature is to build intermediate services that route requests based on certain conditions and provide responses to the consumers, without changing the implementation of the existing applications and services. Similarly, intermediate services can be implemented to extract data from multiple applications and present it to the customers as a single response.

- **API policy enforcement and management**: Different consumers may consume the business services through different channels. Providing access to these customers in a controlled manner is supported by integration platforms with the API management capability. In addition to that, it supports securing and monitoring services.

- **Ecosystem/community management**: If the business is exposing certain business capabilities as APIs, then third-party application developers and partners can build their applications and expand the business to a wider customer base. Integration platforms manage this ecosystem and community of developers via mechanisms such as developer portals, which allow developers to automatically browse APIs and build applications using them.

Now that we understand the capabilities of an integration platform and how it relates to other components of an integration platform, in the next section, we'll discuss how we can put together a solution architecture that describes the concepts we've discussed so far in this chapter.

Solution architecture of an enterprise software system with an integration platform

As we discussed at the beginning of this chapter, most real-world enterprise platforms are brownfield platforms with many different software applications and systems. The microservice architecture based on NATS messaging that we discussed in the previous chapters of this book is only a part of the overall solution. It is essential to understand how all these different components work together to provide the best possible experience to the consumers by helping businesses run efficiently. There are several approaches we can follow to achieve this requirement.

We will be discussing one of the most common approaches we could find in the enterprise architecture, which is the **API-driven integration platform** architecture. In this model, we make most, if not all, the applications and services accessible via APIs and then categorize them based on their functionality and the level of interaction with the end users. The following diagram depicts the high-level view of the API-driven integration architecture:

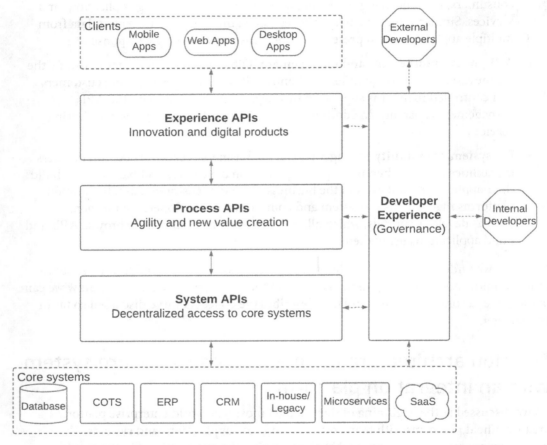

Figure 9.1 – API-driven enterprise platform architecture

The preceding diagram depicts how an API-driven integration platform helps us design an enterprise platform that delivers an omnichannel user experience by using APIs at its core. Let's go through the main components shown here:

- **Clients**: These are the different channels that consumers use to interact with businesses to get services and products. This API-driven approach helps us make these different interactions in a standard manner, rather than supporting client-specific implementations on the enterprise side.

- **External developers**: These are the application developers who build applications to expand the business services to new customers. They do this by integrating the APIs into different business domains that are related to our specific business domain.

- **Experience APIs**: This is the edge of the enterprise platform. It interacts directly with clients and users to provide new products and services. At this layer, we build services that provide what the user requires and apply different Quality of Service (QoS) measures such as security, rate-limiting, caching, and monitoring while acting as proxies to the actual services that are implemented at the lower layers, such as process APIs and system APIs.

- **Process APIs**: This is the intermediate layer, which makes the necessary adjustments and transformations to match the end user demand with the backend services implementation. It does this by processing the data provided by the core services and system APIs. This layer orchestrates multiple systems and applications to provide the best possible response to the client requests that are coming through the experience APIs.

- **System APIs**: This is where the actual data processing happens. Even though this layer is depicted as a separate layer on top of the core systems, most of the core systems provide APIs to access their core services. In such a scenario, there won't be an additional system APIs layer. Instead, the core service becomes the system API.

- **Developer experience**: When all of the systems and applications are exposed as APIs, you must have a centralized repository that allows the developers to find the APIs that have already been developed so that they can reuse those before building new services and APIs. This also becomes useful when exposing the experience APIs to the external developers for third-party application development. This component connects to all the different APIs and provides governance and an experience that is similar to an application store to the internal and external developers.

- **Internal developers**: These are the internal application developers who build applications and services that are consumed by direct customers and internal stakeholders.

- **Core systems**: These are the core enterprise applications that help businesses carry out all the core business operations.

In addition to these components, there will always be other supportive services such as monitoring, infrastructure provisioning, and security integration with this architecture. Let's take a look at how this architecture maps to the components of a real-world integration platform:

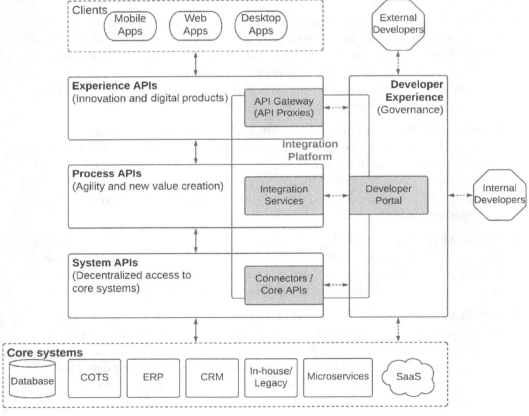

Figure 9.2 – API-driven enterprise mapped to an integration platform

The preceding diagram depicts the components of a typical integration platform that maps to the API-driven enterprise platform. Let's discuss these in detail.

API gateway

Experience APIs need a component that can define new services as and when required by the business. This also allows for security enforcement and other QoS capabilities. An API gateway is a component that is used to expose the standards-based interfaces to the clients so that they can consume the services, regardless of their implementation technology. Typical API gateways allow you to expose REST APIs using Swagger or **OpenAPI Specification** (**OAS**) standards. More recently, API gateways allow you to expose streaming APIs such as WebSocket, as well as asynchronous APIs such as webhooks, WebSub, and Server-Sent Events using the AsyncAPI specification. Enterprises can define their representation of the services using these specifications without exposing the backend implementation details. In addition to that, the gateway enforces certain QoS measures to make sure that services are secured with modern security standards such as OAuth2. They also enable rate-limiting policies to make sure that consumers fairly get access to the services based on the subscription agreements they made with the service provider.

Integration services

At the process APIs layer, we need to execute certain operations such as transformations, orchestrations, and logical executions. An integration platform provides a component that is capable of implementing these types of intermediate tasks using the integration services. Most of these platforms provide a high-level **domain-specific language** (**DSL**) to implement these services so that developers with domain knowledge can easily build services without learning an entire programming language. This integration services component typically provides a set of building blocks that can be assembled using a **graphical user interface** (**GUI**) to build the intermediate services quickly.

Connectors

The systems APIs layer should be capable of directly integrating with various systems that are already used in the enterprise platform. An integration platform provides a set of connectors to integrate with these systems so that the internal complexities of communicating with these systems are not visible to the developers. These connectors provide a higher-level API to the developers so that they can use those connectors in the integration services they build at the process APIs layer.

Developer portal

The API-driven enterprise platform results in a lot of APIs running in the platform for different purposes. Hence, it is essential to have a central repository where API developers and API consumers can browse through the available APIs. That is the functionality of this developer portal component. It provides functionalities such as search, tagging, rating, comments, subscription, try-out, and analytics. This component allows API consumers to share their experiences with the API developers so that developers can improve the existing APIs and create new APIs based on that feedback. It also allows the enterprise platform to selectively expose APIs to external developers. This allows them to build new applications and integrate the provided functionality into those applications to gain more business opportunities for the organization.

At this point, we have a better understanding of how an enterprise platform is designed in a brownfield scenario using an API-drive approach using an integration platform. The next step is to understand how our microservice architecture, along with NATS messaging, fits into this solution. We'll discuss this in the following section.

How do microservices and NATS integrate with the integration platform?

In *Chapter 5*, *Designing a Microservice Architecture with NATS*, we discussed how the outer architecture of the microservice-based approach uses an integration platform to connect to the existing components of the enterprise platform. In addition to that, we discussed how an API gateway acts as the interface to the external consumers in the outer architecture. In this section, we will expand this concept further and come up with an overall solution for an enterprise platform that consists of an integration platform based on an API-driven approach, a microservice-based application, and NATS:

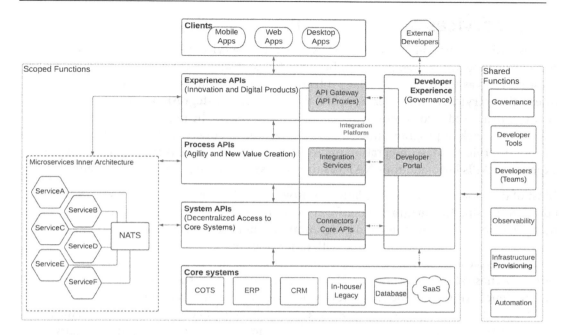

Figure 9.3 – Enterprise architecture with microservices, NATS, and the integration platform

The preceding diagram depicts how a microservice-based application that uses NATS as the messaging layer interacts with an integration platform and other components of an enterprise platform. Let's discuss these newly added components.

The microservice architecture allows you to develop applications for different purposes and all the microservices expose APIs to interact with them. This fits nicely into the API-driven approach that we've discussed so far in this chapter. The preceding diagram depicts that microservices can be implemented at any of the following layers:

- Microservices at the process APIs layer
- Microservices at the system APIs layer
- Microservices at the core systems layer

Let's discuss these different types of microservices in detail and how they interact with other components in the enterprise platform.

Microservices as process APIs

Process APIs are typically implemented as integration services using the capabilities provided by the integration platform. But that does not stop us from implementing these services as microservices. Many integration platforms allow us to develop these integration services as microservices, and these are called **integration microservices**. As we've discussed throughout this book, microservices can be implemented with any technology stack or programming language, given that it adheres to the core principles of the microservices. These integration microservices expose the APIs to integrate with the experience APIs layer and uses NATS as the inter-service communication mechanism.

Most of the modern integration platforms provide mechanisms to integrate with NATS. In addition to using the integration platform to implement process APIs, we can always build these using other technologies and programming languages, such as Go or Java. These microservices can then be integrated with other parts of the system using the integration platform. Typically, these services implement use cases such as data transformation, service orchestration, and conditional routing, to name a few. These services are exposed as experience APIs using the API gateway provided by the integration platform. Given that these integration microservices expose APIs whenever necessary, the API gateways act as a proxy while providing the security, rate-limiting, and monitoring functionality required.

Microservices as system APIs

System APIs provide the low-level business functionality over APIs so that process APIs can build aggregated services that are required by the clients. Microservices are a perfect fit for implementing system APIs since these APIs play an important role in the overall solution. Most of the real-world applications that are built using microservice-based approaches fall into this category. The developers can select their technology stack to develop these services and these services can integrate with the existing core systems, as well as other layers of the integration platform. There is the possibility of using the integration platform to build these system APIs by wrapping the core systems and exposing those systems as APIs.

Microservices as core systems

The core systems act as the source of truth for the business data that stores all the critical business details. These systems are often developed by third-party vendors and used as either cloud services or COTS services. Some of these vendors will support deploying these solutions in a microservice-friendly manner. But there won't be much flexibility for the enterprises to decide on the deployment models. There can be situations where certain business-specific requirements need to be implemented quickly under the purview of the enterprise architects. In such a scenario, the architect can decide on building the application or applications as microservices. These applications need to be implemented with extra attention being paid to key aspects such as performance, availability, failure handling, and scalability. Using a microservice-based approach would be ideal for this use case.

Microservices and the integration platform

As we discussed in the previous section, a microservice-based approach can be used to develop different types of services within the enterprise platform. Once these microservices have been implemented, these services need to interact with the other parts of the platform. If these services support standard interfaces such as REST, then microservices can directly integrate with them without going through the integration platform. But most of these systems have proprietary protocols and messaging formats that are not easy to scale so that they directly integrate with microservices. In such cases, microservices will use the integration platform as the intermediate component that facilitates communication. If these internal systems support NATS messaging, then microservices can communicate with those services via the NATS messaging layer.

In addition to that, the integration platform allows the microservices to be exposed as managed APIs using the API gateway component at the experience APIs layer. The developer portal component registers the microservices as APIs in the portal so that internal and external developers can find these services and reuse them for their applications and services. Shared functions such as observability, automation, infrastructure provisioning, and development-related activities are all integrated as common services for both the integration platform and the microservice architecture.

At this point, we have a better understanding of how microservices, NATS, and integration platforms operate within an enterprise platform. Now, let's go ahead and build such a platform with a few sample services to strengthen our understanding of these concepts. For this, we will use the OPD application that we designed in *Chapter 6, A Practical Example of Microservices with NATS*.

How to build an enterprise platform with microservices, NATS, and the integration platform

Even though we've discussed multiple layers of the API-driven enterprise platform approach, not all these layers are used in all practical scenarios. Depending on the complexity of the application, different layers are introduced. For a simple application such as our OPD application, we don't need all four layers that we discussed in the previous section. The following diagram depicts how our OPD application fits into the API-driven approach we discussed previously:

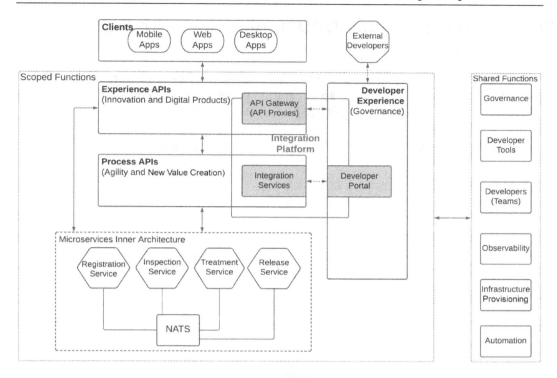

Figure 9.4 – The OPD application with the integration platform

In the preceding diagram, we are implementing our business logic as microservices and using NATS to communicate within these services. In addition to that, we are using an integration services layer to implement certain functionalities that involve interacting with more than one service or modifying the responses that are coming from these services. Also, we are using an API gateway to expose these services to external consumers via an OAuth2.0 protocol-protected manner. Finally, all these microservices are available in the developer portal so that internal and external application developers can see the available services and functionalities these services have to offer, without the need to contact individual microservices teams. Let's go ahead and set up an environment to demonstrate the architecture depicted in the preceding diagram through a real-world example using the WSO2 integration platform.

Setting up the sample OPD application with WSO2 integration platform

In this section, we are going to set up a simple API-driven enterprise platform that consists of microservices, NATS, and an integration platform. We will be using the OPD application code that we developed in *Chapter 6, A Practical Example of Microservices with NATS*, a single-node NATS server, and WSO2 API Manager 4.0 as the integration platform. The following diagram depicts how all of these components are placed architecturally:

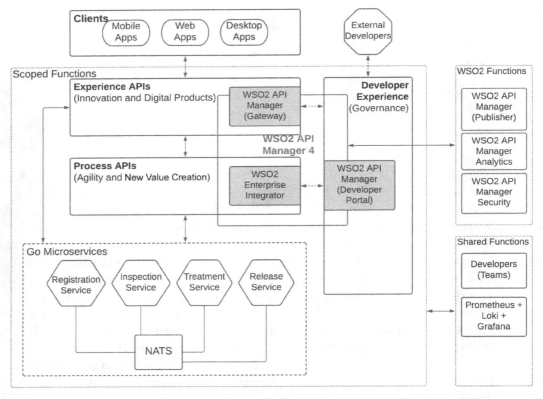

Figure 9.5 – The OPD application with NATS and WSO2 API Manager

The preceding diagram is an extension of *Figure 9.4* and shows the components that are brought into the architecture by WSO2 API Manager 4. Let's describe these additional components in detail:

- **WSO2 API Manager Gateway**: This is where the security and rate-limiting policies are enforced so that microservices are secured with OAuth2-based security. The services can be exposed as standard REST APIs using Swagger or OpenAPI Specification definitions.

- **WSO2 MicroIntegrator**: This is the integration component of the WSO2 API Manager platform. It allows us to do various message processing and orchestration tasks. It uses a high-level XML-based **domain-specific language** (DSL) called **Synapse** to develop these integrations.

- **WSO2 API Manager Developer Portal**: This is where all the microservices are published as discoverable APIs. The consumers of these APIs (typically, the application developers) can find the APIs based on their needs using the mechanisms provided in this tool.

- **WSO2 API Publisher**: This is an interface provided by the platform to develop APIs from the microservices so that those APIs are protected with proper security measures and are also visible in the developer portal. This creates the proxies that are used to enforce the **QoS** measures on microservices at the API gateway. This also acts as the governance layer for APIs, where it can be used to manage the life cycle states of the APIs and other aspects such as API visibility and control.

- **WSO2 API Analytics**: This is the business analytics of the API usage within the platform, and it is provided as a cloud service by WSO2. It provides business insights to the users, enabling them to make certain business decisions when managing the APIs.

- **WSO2 API Security**: This component is responsible for enforcing OAuth2-based security at the gateway, as well as rate-limiting and throttling the requests based on predefined policies.

- **Observability with Prometheus, Loki, and Grafana**: We will use these tools to monitor the microservices as well as the WSO2 API Manager platform for troubleshooting purposes.

Now that we have a better understanding of what we are going to build, let's go ahead and configure the environment.

Setting up the NATS server

The first step is to set up the NATS server so that microservices can communicate with each other. For this sample environment, we are going to use the simplest form of the NATS server, without any additional security or clustering, since we have covered those concepts in detail in the previous sections of this book. You can go ahead and start the NATS server in the default configuration with the following command:

```
$ nats-server -m 8222 -D
```

This command will start the NATS server with client connection port 4222 and monitoring port 8222. The next step is to start the microservices of the OPD application.

Setting up microservices

In this example use case, we are going to use the same code we developed in *Chapter 6, A Practical Example of Microservices with NATS*. This code has been copied into the / chapter9 directory since certain improvements need to be made by you. First, you need to download the source code in this book's GitHub repository and go inside the / chapter9 directory. Then, you can go ahead start the four microservices shown in the following subsections. You can find the same instructions in the README file of the same directory for further reference.

Starting the registration service

Open a new terminal window and execute the following commands:

```
$ cd registration-service
$ go run cmd/main.go -dbName opd_data -dbUser root -dbPassword
Root@1985
```

You should see the following log entry once the service has been started:

```
2021/07/11 13:56:12 Starting NATS Microservices OPD Sample
- Registration Service version 0.1.0 2021/07/11 13:56:12
Listening for HTTP requests on 0.0.0.0:9090
```

In the NATS server log, you should also observe that the registration service client has connected to it.

Starting the inspection service

Open another terminal window and start the inspection service by executing the following commands:

```
$ cd inspection-service
$ go run cmd/main.go -dbName opd_data -dbUser root -dbPassword
Root@1985
```

Once the service has been started, you should see a log entry similar to the following one in the terminal:

```
2021/07/11 13:56:18 Starting NATS Microservices OPD Sample -
Inspection Service version 0.1.0 2021/07/11 13:56:18 Listening
for HTTP requests on 0.0.0.0:9091
```

Additionally, in the NATS server log, you should observe that the inspection service client is connected to the server.

Starting the treatment service

Open yet another terminal window and execute the following commands to start the treatment service:

```
$ cd treatment-service
$ go run cmd/main.go -dbName opd_data -dbUser root -dbPassword
Root@1985
```

The service will start and a log entry similar to the following one will be printed in the terminal:

```
2021/07/11 13:56:26 Starting NATS Microservices OPD Sample -
Treatment Service version 0.1.0 2021/07/11 13:56:26 Listening
for HTTP requests on 0.0.0.0:9092
```

The NATS server will also print the logs related to the connection of the treatment service to the NATS server.

Starting the release service

You can start the release service in another terminal window by executing the following command:

```
$ cd release-service
$ go run cmd/main.go -dbName opd_data -dbUser root -dbPassword
Root@1985
```

These commands start the release service and a log entry, like the one shown here, that gets printed in the terminal window:

```
2021/07/11 13:54:25 Starting NATS Microservices OPD Sample -
Release Service version 0.1.0 2021/07/11 13:54:25 Listening for
HTTP requests on 0.0.0.0:9093
```

The NATS server will also print the logs related to the connection that's been made by the release service client to the server.

At this point, we have the microservices and the NATS server ready to go. Now, let's go ahead and start WSO2 API Manager so that we can expose these microservices as secured APIs to external clients.

Setting up WSO2 API Manager 4

You can download WSO2 API Manager by visiting the following download link: https://wso2.com/api-manager/#. You can choose your preferred option to install it. We are going to use the binary version, which you can download by selecting the **Zip archive** option on this page. It will download the archived binary, which you can extract to a selected location and start using the product easily. You can refer to the installation instructions available at https://apim.docs.wso2.com/en/latest/install-and-setup/install/installing-the-product/installing-api-m-runtime/ for more information.

Once you have set the JAVA_HOME environment variable and the API Manager server is ready to start, you can go ahead and start the server by executing the following command:

```
$ sh api-manager.sh
```

This command starts an all-in-one instance of API Manager that contains the following components:

- WSO2 API Gateway
- WSO2 Developer Portal

- WSO2 API Publisher
- WSO2 API Security

The analytics component will run as a cloud service, while the integration service is a separate component that we need to download from `https://wso2.com/integration/#` and install separately.

The following URLs are available to access the different components of WSO2 API Manager:

- `https://localhost:9443/publisher`: This URL is used to create API proxies, configure security and other QoS, and govern the API life cycle.

- `https://localhost:9443/devportal`: This URL is used to browse through the available APIs and try out the functionality of each API.

- `https://localhost:9443/admin`: This is the administrative interface and is used to configure advanced settings such as API categories, workflows, and users and permissions.

The next step is to create API proxies for our four microservices in the WSO2 API Manager platform so that we can expose these services to external consumers securely and through the developer portal.

Creating API proxies and publishing them on the portal

The following four microservices have been exposed in the URLs mentioned here:

- **Registration Service**: `http://localhost:9090/opd/patient/`
- **Inspection Service**: `http://localhost:9091/opd/inspection/`
- **Treatment Service**: `http://localhost:9092/opd/treatment/`
- **Release Service**: `http://localhost:9093/opd/release`

We are going to create API proxies for all of these services using WSO2 API Manager, as mentioned in the following steps:

1. Let's go into the WSO2 API Manager publisher portal by `going to https://localhost:9443/publisher` and creating an API for the registration service, as mentioned in the following documentation link: `https://apim.docs.wso2.com/en/latest/design/create-api/create-rest-api/create-a-rest-api/`.

    When you create the API, make sure that you use the values shown in the following table:

Name	PatientRegistration
Context	`/registration`
Version	`1.0.0`
Endpoint	`http://localhost:9090/opd/patient/`

Table 9.1 – API creation details

2. Go ahead and click on the **Create & Publish** button to create the API proxy and publish it to the API developer portal.

    What we just did was create an API proxy for our registration microservice and publish that to the developer portal. This allows internal and external developers to find the registration service and use it with a valid subscription and a token.

3. You can try out this API by accessing the developer portal at `https://localhost:9443/devportal`.

    You should see the newly created API, called **PatientRegistration**, and a link to **TRY OUT** this API.

4. Go ahead and try out the API by clicking on the **TRY OUT** button. This will subscribe the default application to this API and generate a token so that you can access the API with a valid token.

5. In the **TRY OUT** interface, you can click on the **GET TEST KEY** button, then select the **GET** method at the bottom section, and then click on the **TRY IT OUT** and **EXECUTE** buttons to get the response from the registration service. This should provide the default response, which is similar to the following:

```
Registration Service v0.1.0
```

Here, we have only tried out the / or `base` context of the service. But the actual context that we are going to use in the real world is not the base context. For the registration service, we need to use the `/register` context with the `POST` method to register a patient to the system.

6. To do that, let's use a `curl` command with the generated token. Make sure that you replace the `<Token>` section with the actual token that was generated in the WSO2 API developer portal:

```
$ curl -X POST "https://localhost:8243/
registration/1.0.0/register" -H "accept: */*" -H
"Authorization: Bearer <Token>" -d '{"full_name":"john
doe","address":"22, sw19, london","id":500, "sex":"male",
"phone":123456789}' -k
```

If the request is successful, you should get a response similar to the following:

```
{"id":500,"token":1}
```

At this point, the registration service has been exposed as an OAuth2 secured API over the API gateway and published to the developer portal. Now, let's go ahead and create the API proxies for the other three services by following the same instructions that we followed for the registration service. This has been left as an exercise for you so that you become familiar with the WSO2 API Manager tool.

Once you have created all the services as API proxies within the WSO2 API Manager platform, you could try out the end-to-end use case by accessing the services with secured JWT tokens.

The component that is used to implement process APIs in the WSO2 integration platform is **WSO2 Enterprise Integrator** (**Micro Integrator**). Let's go ahead and set up that and implement a sample process API to showcase the overall status of a patient. To do that, we will call the registration, inspection, and treatment services from this intermediate service and then respond with an aggregated response.

Setting up WSO2 Micro Integrator

You can download the latest version of WSO2 Micro Integrator from `https://wso2.com/integration/#`. At the time of writing this book, the latest version is 7.1.0, so we will use that for our example. Go ahead and choose **Zip archive** from the download options. Then, you can extract this archive and install WSO2 Micro Integrator using the instructions mentioned at `https://ei.docs.wso2.com/en/latest/micro-integrator/setup/installation/install_in_vm_binary/`.

Once you have WSO2 Micro Integrator installed in your setup, you can go ahead and start the server with the following command:

```
$ sh micro-integrator.sh
```

In addition to the runtime, you need an IDE to develop integrated services with WSO2 Micro Integrator. You can download WSO2 Integration Studio from https://wso2.com/integration/integration-studio/.

You can install the binary in your environment and create a new integration project to implement our use case. For this example, we are going to implement a service orchestration scenario by calling three services and aggregating the responses from these services so that we can provide a singular view of the patient to the end user. The following sample configuration file can be used to achieve this requirement:

```xml
<?xml version="1.0" encoding="UTF-8"?>
<api context="/patient" name="PatientData"
  xmlns="http://ws.apache.org/ns/synapse">
    <resource methods="GET" uri-template="/{id}">
        <inSequence>
            <property description="patient-id"
             expression="$ctx:uri.var.id" name="patient_id"
              scope="default" type="STRING"/>
            <log>
                <property expression="$ctx:patient_id"
                 name="STATUS"/>
            </log>
            <call>
                <endpoint>
                    <http method="get" uri-template=
                     "http://localhost:9090/opd/patient
                      /view/{uri.var.id}">
                    </http>
                </endpoint>
            </call>
            <property expression="json-eval($.)"
             name="patient_response"
              scope="default" type="STRING"/>
```

In the first part of the code, we are storing the patient ID parameter, which comes as a path parameter in the request URL to a property (similar to a variable), and then prints that into the log file. Then, we call the patient registration service with this patient ID to get the details of this patient. The response from this service is stored in another property called `patient_response` so that we can use that later when responding with the aggregated response:

```
<call>
    <endpoint>
        <http method="get" uri-template
        ="http://localhost:9091/opd/
            inspection/history/{uri.var.id}">
        </http>
    </endpoint>
</call>
<property expression="json-eval($.)"
  name="history_response"
    scope="default" type="STRING"/>
<call>
    <endpoint>
        <http method="get" uri-
            template="http://localhost:9092/opd/
            treatment/history/{uri.var.id}">
        </http>
    </endpoint>
</call>
<property expression="json-eval($.)"
  name="treatment_response"
    scope="default" type="STRING"/>
```

In the aforementioned code segment, we are calling the inspection service and the treatment service with the patient ID to get the relevant details. Also, we are storing these responses in the `history_response` and `treatment_response` properties, respectively:

```
<payloadFactory media-type="json">
    <format>{"patient_summary" : {"details" :
        "$1", "inspections" : "$2", "treatments"
        : "$3"}}
        </format>
    <args>
        <arg evaluator="xml"
            expression="$ctx:patient_response"/>
        <arg evaluator="xml"
            expression="$ctx:history_response"/>
        <arg evaluator="xml"
            expression="$ctx:treatment_response"/>
    </args>
</payloadFactory>
<respond/>
</inSequence>
<outSequence/>
<faultSequence/>
</resource>
</api>
```

In the final code segment, we are creating a new payload using the values stored in the properties in the previous sections and then sending that aggregated payload to the client.

The preceding integration service is implemented using the synapse language, which is the DSL that's used to build integrations for WSO2 Micro Integrator. You can find the deployable archive file (`OPDProjectCompositeExporter_1.0.0-SNAPSHOT.car`) at `https://github.com/PacktPublishing/Designing-Microservices-Platforms-with-NATS/tree/main/chapter9`.

You can copy this file to the Integrator runtime under the `/micro-integrator/repository/deployment/server/carbon-apps` directory. This will automatically deploy the integration service to the runtime. Now, we can call this service to check its functionality by executing the following command:

```
$ curl localhost:8290/patient/500
```

This command should give an aggregated response from the three services, similar to the following:

```
{"patient_summary" : {"details" : "{"text":"{\"full_
name\":\"chanaka fernando\",\"address\":\"44, sw19,
london\",\"id\":500,\"sex\":\"male\",\"phone\":222222222}\n"}",
"inspections" : "{"text":"[{\"id\":500,\"time\":\"2021/07/12
10:21 AM\",\"observations\":\"ABC
Syncrome\",\"medication\":\"XYZ x 3\",\"tests\":\"FBT,
UC\",\"notes\":\"possible Covid-19\"},{\"id\":500,\"t
ime\":\"2021/07/12 10:21 AM\",\"observations\":\"ABC
Syncrome\",\"medication\":\"XYZ x 3\",\"tests\":\"FBT,
UC\",\"notes\":\"possible Covid-19\"}]\n"}", "treatments"
: "{"text":"[{\"id\":500,\"time\":\"2021 07 12 4:35
PM\",\"notes\":\"low fever\"}]\n"}"}}
```

We can expose this aggregation service as a protected API using the API gateway by creating an API from this service. That task is left as an exercise for you.

The next step is to set up the observability for the solution. Let's go ahead and do that using Prometheus, Loki, and Grafana.

Setting up observability

Having separate components for different functional requirements is one of the fundamental concepts of any microservice architecture. When it comes to troubleshooting issues and recovering from failures, it is always useful to have a central place to look into. That is what we discussed in *Chapter 8, Observability with NATS in a Microservice Architecture*. There, we discussed how to configure observability features in the sample microservices and the NATS server. You can refer to that chapter and add observability capabilities to the OPD application microservices that we used in this chapter. We will leave that as an exercise for you.

In this section, we are going to discuss how to enable observability for the integration platform so that all the components of the platform, including microservices and NATS servers, are monitored through a centralized platform.

Configuring WSO2 API Manager for observability

WSO2 API Manager uses logs to print useful information related to the events occurring in the runtime. It provides separate logs to identify different types of activities such as error logs, access logs, audit logs, and generic server logs. We can configure the log monitoring solution we used in *Chapter 8, Observability with NATS in a Microservice Architecture*, with Loki, Promtail, and Grafana. We can add the following configurations to the Promtail configuration file:

```
- job_name: wso2-apim-carbon
  static_configs:
  - targets:
    - localhost
  labels:
    job: wso2-apim-carbon
    __path__: /Users/chanakaf/Documents/source-code/
      certs/wso2am-4.0.0/repository/logs/wso2carbon.log
- job_name: wso2-apim-audit
  static_configs:
    - targets:
      - localhost
    labels:
      job: wso2-apim-audit
      __path__: /Users/chanakaf/Documents/source
        -code/certs/wso2am-4.0.0/repository/logs/audit.log
- job_name: wso2-apim-errors
  static_configs:
    - targets:
      - localhost
    labels:
      job: wso2-apim-errors
      __path__: /Users/chanakaf/Documents/source
        -code/certs/wso2am-4.0.0/repository/logs/wso2
          -apigw-errors.log
```

In the aforementioned configuration, we have added three log files related to server logs, audit logs, and error logs in WSO2 APIM to our log monitoring system. If required, you can add other log files based on the requirements. If you restart the Promtail service with the updated configuration file, you should be able to view the WSO2 API Manager logs from Grafana.

Configuring WSO2 Micro Integrator for observability

WSO2 Micro Integrator also provides a set of log files, similar to WSO2 API Manager, which we configured in the preceding section. You can follow a similar approach to configure log monitoring for WSO2 Micro Integrator. In addition to that, it also provides additional metrics information that we can listen to via Prometheus and Grafana. To enable that, we can add the following configuration to the `deployment.toml` file in the Micro Integrator server:

```
[[synapse_handlers]]
name="CustomObservabilityHandler" class="org.wso2.micro.
integrator.observability.metric.handler.MetricHandler"
```

Once this configuration has been added, the Micro Integrator server can be started with the following command to enable observability metrics:

```
$ sh micro-integrator.sh -DenablePrometheusApi=true
```

Then, you can add the following configuration to the Prometheus configuration file and restart the Prometheus server:

```
- job_name: esb_stats
  metrics_path: /metric-service/metrics
  static_configs:
    - targets: ['localhost:9201']
```

Finally, you can add the Grafana dashboard, which is developed by the WSO2 team, to monitor the behavior of WSO2 Micro Integrator. You can get the dashboard configurations from `https://grafana.com/grafana/dashboards/12887` and add the relevant dashboard to your Grafana server.

At this point, we have our enterprise platform with microservices, the NATS server, and the integration platform with security and observability enabled. This is a sample implementation that we used to showcase the integration of the aforementioned components. You can apply these integration of the preceding mentioned concepts. when you are building real-world enterprise applications.

Summary

In this chapter, we discussed how typical enterprise software systems are built using different types of applications, as well as how the integration platforms help such systems build systems that interact with these different systems. We talked about the brownfield enterprise architecture and how that is being transformed into an API-driven enterprise platform using an integration platform. Then, we looked at how microservices fit into this architecture and work shoulder to shoulder with the integration platform.

Finally, we implemented a sample enterprise platform with our OPD application, the NATS server, and the WSO2 integration platform. By reading this chapter, you learned the concepts of enterprise software systems and how disparate systems work together to achieve common business goals by using an API-driven approach with an integration platform. This chapter also helped you understand how microservices coexist with integration platforms in a typical enterprise system. Now, you are ready to build your next enterprise software project with microservices while keeping the existing services and applications intact.

In the next chapter, which is the final chapter of this book, we are going to discuss the future direction of microservices and NATS and the new developments happening in the enterprise application domain.

10
Future of the Microservice Architecture and NATS

So far, we have looked at the concepts of microservices and the NATS messaging technology. We learned how to use those concepts in real-world applications with a few practical examples. In this final chapter, we are going to discuss the future direction of the microservice architecture and NATS. Both the microservice architecture and NATS are recent technologies with many areas of improvement. The more we work on these technologies, the more we will understand the value of them, as well as their complexities.

In this chapter, we will discuss the challenges imposed by the approach we presented in this book regarding building a microservice-based enterprise platform with NATS. We will understand what measures can be taken to face those challenges and conquer them. We will also understand the several other approaches taken by developers and architects to build microservice-based applications. We will look at some of these new developments and how those developments can shape our proposed architecture.

The following topics will be covering in this chapter:

- Challenges of the proposed microservice architecture with NATS
- What are the new developments in the microservice architecture?
- What are the new developments in the NATS messaging technology?
- How can a microservice architecture with NATS be improved?

By the end of this chapter, you will be able to design future-proof distributed applications using the microservice architecture and NATS.

Technical requirements

In this chapter, we will be configuring the NATS server with JetStream and showcasing its functionality. The following software components need to be installed if you wish to try out the examples mentioned in this chapter:

- The NATS server
- The NATS **command-line interface (CLI)** tools

Challenges of the proposed microservice architecture with NATS

In the previous chapters, we went through practical examples of implementing microservices with their scope and data while sharing information in a secured manner through the NATS messaging technology. In this section, we are going to discuss some of the areas that we need to pay extra attention to when building applications with the proposed architecture.

Message delivery guarantees and error handling

NATS is designed to offer simplicity and performance to users. These design principles can sometimes make things challenging for certain applications. The NATS core technology offers an *at- most -once* message delivery guarantee. In this mode, a message is delivered to a subscriber no more than one time. So, there is a possibility of message loss here. This mode of operation is suited for many practical applications, including the scenarios mentioned here:

- Applications that use a request-reply-based communication for information sharing. In these applications, the client can handle the failure scenarios by configuring timeouts and error codes. The NATS server can operate with the at-most-once strategy.

- The expected message consumers are designed to be available all the time with auto-recovery.

- Messages are delivered at a faster rate so that a lost message can be tolerated with the subsequent messages.

The only downside of this approach is that messages can be lost due to the unavailability of the consumers. A solution to dealing with message loss is to use a sequence ID for messages. By using these sequences, the consumers can detect any message losses and request the message publisher to resend. These sequence IDs can be included within the message payload or can be added to the subject field. Then, consumers can subscribe to the entire sequence using a wildcard subscription.

As an example, a patient registration service can send patient registration messages in a set of subjects with the `patient.registration.1`, `patient.registration.2`, and `patient.registration.3` sequence IDs. Then, the consumer of these messages can subscribe to the `patient.registration.*` subject and listen to all the messages. The consumer can parse the subject to detect the sequence number and determine if there is any message loss and ask for missing messages.

But not all applications can operate with the default at-most-once message delivery guarantee mode. In such scenarios, NATS offers the at least once delivery guarantee through the JetStream feature, which can be enabled with a flag when starting the NATS server. In this mode, a message is delivered to at least one of the subscribers and it avoids message loss in cases where the subscribers are not available or faulty. This can be used with scenarios such as the following:

- When consumers and producers are highly decoupled, and they can come online independently at different times.

- When consumers and producers operate at different message processing speeds.

- When the consumers can go offline due to various issues, yet the messages that are published on a given subject that are listened to by the consumer need to be delivered.

- When a historical record of a message stream is required by consumers. The message replay feature can be used for this purpose.

Most of the practical applications can be implemented with the core NATS technology with the at-most-once delivery mode, which provides the best performance in terms of message exchange through the system. However, JetStream does provide better guarantees, even though it costs more resources and reduces the overall performance.

In addition to these options, NATS allows messages to be replicated across clustered nodes so that the failure of one or more NATS servers will not affect how messages are distributed to consumers. Also, publishers and consumers can connect to different NATS servers, given that they have formed a common NATS cluster.

Security

This is one of the most complicated topics in the world of microservices. In *Chapter 7, Securing a Microservices Architecture with NATS*, we discussed several options for securing microservices for external interactions (North-South traffic), as well as for internal interactions (East-West traffic). In most cases, external interactions are critical since they impact the user experience as well as the security of business services and data directly.

Based on what we are seeing in the technology world, most people are moving toward using OAuth2.0-based security for securing external interactions. With more programming languages, frameworks, and tools providing support for OAuth2.0-based security, we can easily implement it with one of many options available with those tools. An easier way to implement OAuth2.0-based security for external interactions is by using an API gateway that provides built-in capabilities to enable it.

Even though there is a clear agreement in the technology world on using OAuth2.0 for North-South traffic protection, there is no such agreement for protecting East-West traffic. There are good reasons for this lack of agreement. Since securing East-West traffic involves systems that are within the boundaries of the enterprise secured network, some people (especially developers) argue that having **transport layer security** (**TLS**) is sufficient and that, if at all, using the **mutual TLS** (**mTLS**) is more than enough. That is true for the most part and you can follow that approach if you can convince your security experts regarding that approach. We discussed and showcased examples of enabling TLS for interservice communication in *Chapter 7, Securing a Microservices Architecture with NATS*.

But there is another school of thought that comes mostly from security experts that this interservice communication also needs to be protected with OAuth2.0-based security. This is a valid argument and in applications where it requires extreme security controls, you can achieve this using the OAuth2.0-based security features available in the NATS server. In the current implementation of the NATS server, these tokens are managed and maintained by the NATS ecosystem, and it does not have any options to integrate with the OAuth2.0 security services provided by the identity servers that are used for North-South traffic protection.

If there is a need to secure end-to-end communication from the client to the gateway and the microservices, then additional work needs to be done at the microservices layer. Each service should extract the user details from the OAuth2.0 token (JSON Web Token) and use those details to get access to the NATS server that is protected with OAuth2.0-based security. There should be a mapping between the users who are consuming the services via the API gateway and the users who have been configured in the NATS server with their respective roles.

Deployment and automation

So far, we've discussed using containers and container orchestration systems to manage the deployment of the microservice architecture. It is up to the developers and architects to decide on the deployment technology for your microservice-based applications at the start of the project. This allows the teams to design and build their deployment pipelines and automation flow accordingly. Even though the best practice is to deploy microservices in containers backed by container orchestration technology, it does not prevent you from using **virtual machines** (**VMs**) to deploy microservices since most VM-based platforms allow a lot of tooling around automation that is required to build microservices.

Furthermore, you also need to take care of deploying the NATS servers separately from the microservices. In most use cases, you can live with a three-node NATS server cluster. NATS provides resources to deploy these servers in physical hardware, virtual machines, containers, and container orchestration systems such as Kubernetes.

The other important aspect is to look at how you are going to manage the infrastructure. If you have a team with enough resources to manage the infrastructure, you can deploy the container and orchestration technologies by yourself; this will provide more flexibility. But at the same time, it will add more work to your operations team as they would need to pay extra attention to keeping those deployments available for applications.

Another option you can consider here is to go with a managed container and orchestration technology offered by a cloud service provider. Most of the leading cloud vendors such as AWS, Google, and Microsoft Azure provide these infrastructure capabilities as managed services. These solutions provide excellent support for automation and infrastructure monitoring, along with security for the infrastructure.

If you are to build an end-to-end application release workflow, you need to set up several tools and technologies and maintain those tools by yourself to make sure nothing goes wrong during the production releases. Although it is not an impossible task, it would require extra effort to manage these tools, which could adversely impact your application development and delivery dates. Instead, you can use cloud services for these tasks. Most of the cloud infrastructure providers have these tools integrated into their platforms so that we can use them easily with a simple configuration. NATS is also offered as a cloud-hosted, managed service to be used in a full **Software as a Service (SaaS)** manner.

What are the new developments in the microservice architecture?

As we've discussed throughout this book, the microservice architecture is an evolving architectural pattern that's used to build distributed applications. Given that the microservice architecture is used across different industries to build different applications across the globe, new concepts, patterns, and approaches have evolved from those practical implementations. In this section, we are going to touch upon three interesting developments that are happening in the microservices domain at the time of writing this book.

Service mesh for interservice communication

The primary reason for authoring this book was to provide an approach to handle interservice communication within a microservice architecture. We presented the idea of using an intermediate message broker (NATS) to satisfy the requirements of interservice communication. However, there is another alternative approach that is evolving in the microservices arena for interservice communication called the **service mesh**. As its name suggests, it is a mesh topology, which we discussed in *Chapter 1, Introduction to the Microservice Architecture?*. It is the topology that is used to allow services to communicate with each other. In a way, it is going back to the complex point-to-point integration pattern that was used in the early days of integration projects. But the difference here is that the control and the formulation of the mesh are taken out of the implementation of the services using the data and control plane of the service mesh, as depicted in the following diagram:

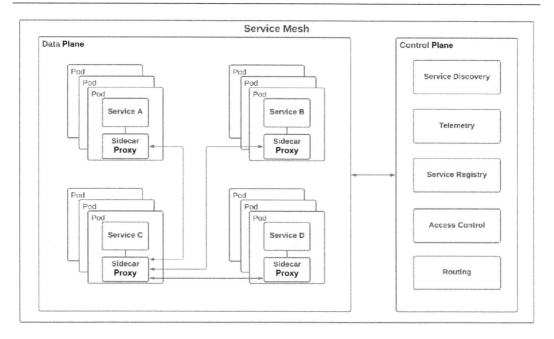

Figure 10.1 – Service mesh for microservices

The primary objective of the service mesh is to allow services to communicate with each other. A service mesh consists of two main components, as depicted in the preceding diagram. These components are as follows:

- **Data plane**: The data plane is the layer that carries data across services and handles the communication aspects. Each service has a sidecar proxy that is responsible for communicating with other services and handling all the network-related complexities. The service will connect to the proxy and proxy route the traffic to the other services (egress), as well as toward the service itself (ingress).

  Both the sidecar and the service run on the same host and connect over localhost. In a Kubernetes-based deployment, both components run within the same pod. In a VM-based deployment, both components run on the same VM.

- **Control plane**: The control plane is used to configure the service mesh. When the services are connected using the data plane, these connections need to be configured with routing, security, and monitoring. These configurations are applied through the control plane. The data plane accepts these configurations and forms the service mesh so that services can communicate with each other. It also allows you to configure resiliency features such as timeouts, retry, and circuit breakers on the sidecar proxies that are running on the data plane.

Istio is the most popular service mesh technology available at the time of writing this book. One of the main challenges with service mesh is the complexity it brings to the overall architecture and the tight coupling nature between services due to its point-to-point connectivity.

Saga pattern and managing data

One of the most challenging aspects of designing and implementing microservice-based applications is handling data. In our OPD application, we used the approach of having a database per service, and any data sharing was done through the events that were sent through the NATS server. This pattern is sufficient and works well for most use cases. But other use cases require sharing data across services in a transactional manner. This means that the data that's processed within one service can impact the data that's processed within another service or several services. That is the main reason to use the saga pattern to develop microservices.

If we take a simple order processing application that takes an order and delivers it to the customer, this application can be implemented as a set of microservices. Let's assume that there are four steps in the order processing application, as follows:

1. Order placement
2. Payment processing
3. Inventory update
4. Order delivery

If we think about the end-to-end process, this needs to be done in a transactional manner and if any of the steps fail, the entire process needs to be reversed. The saga pattern allows you to execute this kind of process as a sequence of steps in a transactional manner. Every step in the process happens through a local transaction that is executed by that service, and it must provide a compensating option to reverse the local transaction. This compensating option is used to reverse the local transaction in case of a failure at any step along the way. This compensating service must be idempotent (executing it repeatedly should provide the same result) and retriable. The following diagram depicts the execution flow of the order processing service:

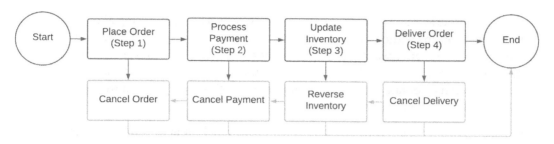

Figure 10.2 – Order processing application execution flow

The preceding diagram depicts the transactional steps of successfully executing the order processing application, as well as the compensating steps that are used to reverse its execution. The actual implementation of the saga pattern can be done using one of the following options:

- **Choreography-based execution**: In this approach, each service can execute the local transaction (step) and pass the results (pass or fail) to the other services as events. Then, there should be a transaction manager that is embedded in each service, or as a standalone component, that checks the results of the local transactions and executes compensating transactions if required.

- **Orchestration-based execution**: In this approach, there is a central coordinator that is responsible for executing the steps and the compensating transactions based on the results. This is much cleaner to implement and suits most of the practical scenarios.

The saga pattern is a useful pattern to use when there is a need to implement transactional data processing within an application. Having said that, it is always recommended to avoid these kinds of complex patterns when implementing microservices as much as possible.

Serverless computing and microservices

Building applications based on the microservice architecture comes with a heavy workload under the hood. One of the major challenges of microservice architecture-based applications is the complexity of managing the underlying infrastructure. We discussed how technologies such as containers and container orchestration platforms offered by cloud computing vendors can help in building microservices without the need to worry much about the underlying infrastructure (especially the servers and virtual machines). Even with such managed container services, there is quite a bit of overhead associated with those technologies since we need to learn how to deal with containers, orchestration, and managing the platform according to our needs.

This is where **serverless computing** comes in handy. Serverless computing allows us to run our applications without worrying about the underlying infrastructure at all. In serverless computing, once we submit the application code to the serverless platform, it will take care of running the application within a suitable infrastructure. It can be a VM or a container or even a container orchestration platform. We do not need to worry about the details of the underlying infrastructure.

One key aspect of serverless computing is that it is designed for applications that can shut down when they're not in use and restart when there is a need. We can specify the deployment parameters such as the number of replicas, auto-scaling requirements, and triggers that will boot up the application. **Functions as a Service (FaaS)** is a common offering of serverless computing that is offered by most of cloud computing vendors such as Amazon Web Services, Google, and Microsoft. We can use these services to implement our microservices as functions and then execute them in serverless mode. This would reduce the overhead of infrastructure management from the developers. In addition to that, since services are running only when there is demand, the cost associated with the underlying infrastructure can also be reduced with serverless computing.

There is still a lot to be improved on serverless computing services offered by cloud vendors. Not all types of applications can be deployed in these serverless platforms due to limitations such as the following:

- Lack of monitoring and debugging capabilities.

- Latencies due to initial startup delays.

- Not suitable for running long-running applications.

- Managing application state is not easy due to the ephemeral (come-and-go) nature of the services.

These limitations will hopefully be gone with the new versions of these services in the future.

What are the new developments in the NATS messaging technology?

The NATS server implementation has gone through several iterations within the past couple of years. In this book, we have used the latest version that is available at the time of writing, which is NATS 2.2. We have not covered all the features of NATS within this book since a separate book would be required to do that. One major change NATS has made with the version 2.2 release is the introduction of JetStream as the successor of the NATS streaming feature. We'll discuss the JetStream feature of NATS in detail in the following section.

JetStream

Throughout this book, we have discussed how to utilize an event-based messaging approach for interservice communication within a microservice-based application design. We used NATS as the event hub that exchanges events (messages) across services. A sequence of events that adhere to a common schema (format) is called an **event stream**. Unlike processing a single event, processing an event stream requires additional capabilities to maintain the correlation and the history of events. This task of processing a stream of events is called **stream processing**. There are several stream processing tools available in the technology world, including Apache Storm, Apache Flink, Kafka Streams, and WSO2 Siddhi, to name a few.

JetStream is designed for stream processing. It tries to solve some of the problems in the existing stream processing tools while providing better functionality. Most of the existing tools struggle when it comes to horizontal scalability, multi-tenancy, and deployment choices. JetStream tries to address these concerns with its design by building these capabilities as core features of the implementation, rather than adding these features later. In addition to these unique capabilities, JetStream provides the necessary functionality that is missing from the core version of NATS to build a reliable messaging framework that can process individual events, as well as event streams. Here is a list of features supported in JetStream:

- **At least once delivery**: This provides a better message delivery guarantee and makes sure messages are not lost.

- **Exactly once delivery within a window**: This avoids message duplicates and improves application reliability.

- **Persist messages and replay**: This helps applications have publishers and consumers that process messages at different rates and replay messages if a consumer failed to process a given message or a set of messages.

- **Persist streams and replay via consumers**: This allows message streams to be persisted and replayed once the consumers are available to process the streams.
- **Wildcard support for streams**: This allows consumers to listen to streams based on wildcards so that multiple streams can be consumed through a single subscription.

In terms of the runtime, JetStream is a feature that is embedded into the NATS server. It can be enabled by setting a flag (-js) when starting the server. This allows users to have deployments with a mix of JetStream-enabled servers and standard NATS servers so that different use cases are catered to by different servers. If there is a connection to a JetStream-enabled server from any NATS server, consumers can produce and consume streams from that NATS server.

The following command starts the NATS server with JetStream enabled:

```
$nats-server -js
```

Once this command has been executed, in the terminal log, you will be able to observe that the JetStream function started in the NATS server. Now, let's dig deeper into the concepts of JetStream. This will provide us with a proper understanding of what it provides for our future architecture. The following are the two main concepts in JetStream:

- Streams
- Consumers

We'll go through each of these concepts in detail in the following subsections.

Streams

A **stream** is a collection of messages. In JetStream, streams are used to define how messages are stored and how long messages are retained for (retention duration). We can define one or more subjects that will be used to collect messages to the stream. Those messages are then stored in the defined storage option. When publishing messages to the streams, you can do a standard publish, where the messages will be stored without sending any acknowledgment to the client. If you send a message in request/reply mode, the NATS server will reply with an acknowledgment that says the messages have been stored. A full list of configurations that can be set on a stream can be found in the NATS documentation at https://docs.nats.io/jetstream/concepts/streams.

We can try out the JetStream functionality by installing the NATS CLI tools, which can be downloaded and installed from https://github.com/nats-io/natscli/releases/.

Once you have installed the CLI tool, we can go ahead and try out a few examples to improve our understanding of streams in JetStream. Let's assume that we need to create a stream to store messages related to patients so that we can display the number of patients in the OPD unit at a given time. We can define a stream with the following command:

```
$ nats str add patients
? Subjects to consume patients.*
? Storage backend file
? Retention Policy Limits
? Discard Policy Old
? Message count limit -1
? Message size limit -1
? Maximum message age limit 1y
? Maximum individual message size [? for help] (-1) -1
```

Once this command has been executed, we are ready to publish messages to the **patients** stream. This stream has the following characteristics:

- It stores the messages in the filesystem.
- It stores messages, so long as there is enough storage in the filesystem.
- It keeps the messages for 1 year before discarding them.
- It collects messages that are published on subjects starting with patients.

Let's go ahead and publish a couple of messages to see what is happening.

We can use the standard pub/sub-based approach to publish a message to the stream without any acknowledgment using the following command:

```
$ nats pub patients.new {"ID":5}
```

This command publishes the message to the patients stream and it is stored in the filesystem. This is different than the standard NATS server operation since the standard server would discard the message since there are no consumers on this subject. But here, we are storing the message so that consumers can come online and consume the messages at their speed.

If we need to publish the message with an acknowledgment so that we can be sure that the message is stored in the NATS server properly, we can use the following command, which uses the /reply mode to send the message:

```
$ nats req patients.discharge {"ID":3}
```

In this scenario, you should see the response message, as shown in here, in the terminal window that was used to execute the preceding command:

```
11:41:34 Sending request on "patients.discharge"
11:41:34 Received on "_INBOX.iy7Q1VqMd4GFL6637mPfYb.
NXgCeMVGK7200000000000" rtt 476.862µs
{"stream":"patients","seq":2}
```

At this point, we have a proper understanding of streams and how to publish messages to streams using JetStream. Now, let's go ahead and discuss the consumers who can receive these messages.

Consumers

The messages that are published to streams need to be consumed by the consumers before they expire or before they are dropped due to various policies set at the streams such as maximum size, maximum age, or any other consumer-related policies. A stream can have one or more consumers and these consumers can use different configuration options when consuming messages from a stream. These consumers are considered smart consumers since they can control the way the messages are received at their end, rather than giving that control to the NATS server, as in the default NATS server mode.

JetStream supports both pull-based and push-based consumers. In the pull model, the consumer requests the messages from the NATS server based on its requirements. But in the push model, the server will send messages to the consumers as soon as the messages arrive at the server. Consumers need to track their progress of message consumption and configure acknowledgment policies so that messages are not duplicated. You can learn about various consumer-related configurations at `https://docs.nats.io/jetstream/concepts/consumers`.

Let's go ahead and create a consumer for the patients stream that we created in the previous section. We can do this by specifying the consumer configuration like so. Here, we are creating a pull-based consumer:

```
$ nats con add --sample 100
? Consumer name NEW
? Delivery target (empty for Pull Consumers)
? Start policy (all, new, last, subject, 1h, msg sequence) all
? Replay policy instant
? Filter Stream by subject (blank for all) patients.new
? Maximum Allowed Deliveries 20
```

```
? Maximum Acknowledgements Pending 0
? Select a Stream patients
```

In the preceding command, we are creating a consumer with the following configurations:

- It reads messages from the patients stream.
- It is called NEW.
- It reads all the messages from the stream.
- It filters messages based on the patients.new subject.
- It allows redeliveries up to the count of 20 if the acknowledgment is not received.

We can configure many other parameters related to the consumer using the same approach. As you may recall, we published a message to the stream with the patients. new subject in the previous section. Now, let's go ahead and read that message from the stream using this consumer. To do that, we can use the NATS CLI tool and execute the following command:

```
$ nats con next patients NEW
```

Using this command, we are reading the next available message from the consumer called NEW. This command will call the NATS server, read the message from the stream, and produce an output similar to the one shown here in the terminal:

```
[12:05:32] subj: patients.new / tries: 1 / cons seq: 1 / str
seq: 1 / pending: 0
{ID:1}
Acknowledged message
```

If we try to read the message again with the same command, it will throw an error since there are no more messages to be consumed by this consumer.

We can create a push-based consumer with the following command:

```
$ nats con add
? Consumer name OBSERVE
? Delivery target (empty for Pull Consumers) observe.patients
? Delivery Queue Group
? Start policy (all, new, last, subject, 1h, msg sequence) last
? Acknowledgement policy none
? Replay policy instant
```

```
? Filter Stream by subject (blank for all)
? Idle Heartbeat 0s
? Enable Flow Control, ie --flow-control No
? Select a Stream patients
```

In the preceding command, we are creating a consumer with the following configurations:

- It is called OBSERVE.
- It reads messages from the patients stream.
- It starts reading from the last message.
- It delivers messages to the observe.patients subject.
- It replays messages instantly.

Now, our consumer is ready to consume messages using the push model. Let's go ahead and publish a new message to the stream using the following command:

```
$ nats pub patients.new {"ID":9}
```

Now, our consumer is reading messages from the patients stream on all subjects and publishing those messages to the observe.patients subject. Let's go ahead and subscribe to that subject to read the messages that are consumed by our consumers. We can do that by executing the following command:

```
$ nats sub observe.patients
```

You should observe that the message that we just published, as well as the message that we published to the patients.discharge subject in the previous section, is received by the created subscriber. The output should look like this:

```
[#1] Received JetStream message: consumer: patients > OBSERVE /
subject: patients.discharge / delivered: 1 / consumer seq: 1 /
stream seq: 2 / ack: false
{ID:3}
```

```
[#2] Received JetStream message: consumer: patients > OBSERVE /
subject: patients.new / delivered: 1 / consumer seq: 2 / stream
seq: 3 / ack: false
{ID:9}
```

There are several other commands and options available for working with streams and consumers in JetStream. You can learn more about these commands at `https://docs.nats.io/jetstream/jetstream`.

Using NATS to build an observability solution

The solution approach that we discussed in this book can be used to build applications related to different business domains and use cases. The solution we've discussed mainly highlights the usage of NATS for interservice communication in a microservice architecture. In *Chapter 8, Observability with NATS in a Microservice Architecture,* we discussed how an application can write log entries to the files with sample code. In that approach, each microservice writes the log details to the log files in a synchronous manner. This approach can impact the performance of the individual microservices due to the log writing operation happening alongside the business logic's execution. Instead of that, we can build an asynchronous logging solution using NATS by sending the log entries to a separate logger service using `NATS-Server`.

The following diagram depicts the approach of using NATS to build an asynchronous logging solution:

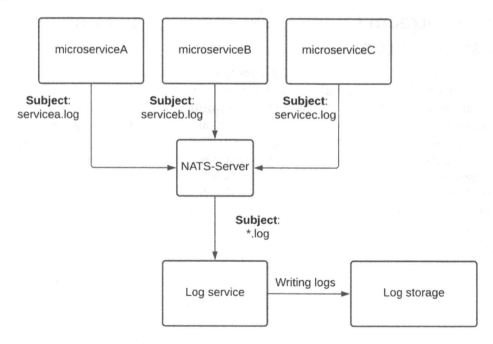

Figure 10.3 – Asynchronous logging solution with NATS

As per the preceding diagram, each microservice publishes log entries as asynchronous messages to the NATS server with subjects representing the microservice's details. In this use case, microservices publish log messages with the subjects `servicea.log`, `serviceb.log`, and `servicec.log`. Then, the log service listens to the log messages using a wildcard subscription with a subject name of `*.log`. Once the messages have been received, the log service can write the log entries to the log storage, without it impacting the performance of the individual microservices. We can use JetStream capabilities to persist log messages if required.

How can a microservice architecture with NATS be improved?

At the beginning of this chapter, we discussed the challenges that are inevitable when designing microservice-based applications with NATS. Then, we discussed some of the new developments happening in the microservice architecture domain and the NATS server technology. In this section, we are going to discuss how we can improve the proposed microservice architecture with NATS using some of these new developments.

Using JetStream for improved message guarantees and stream processing

While designing the OPD application in *Chapter 6, A Practical Example of Microservices with NATS*, we had not considered the possibility of losing messages due to the unavailability of one of the services. But in the real world, when designing an application for a domain such as healthcare, we need to pay extra attention to the message delivery guarantees. This will ensure that all the major events that we share across services such as patient registration, patient inspection, patient treatment, and patient release are delivered to the respective services without any failure. The solution for these types of requirements lies with JetStream. We can easily enable it in the NATS server and make use of streams and consumers to persist these events and deliver them reliably. The following diagram depicts how our OPD application can be improved for message delivery guarantees by using the JetStream feature, along with streams:

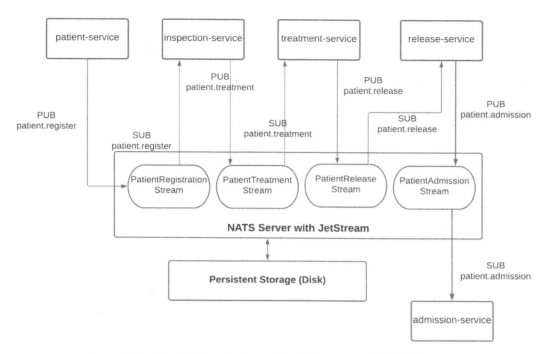

Figure 10.4 – The OPD application with a JetStream-enabled NATS server

The preceding diagram depicts how we can improve the overall design by using a JetStream-enabled NATS server, along with the streams that have been defined for specific message exchanges. As an example, we can define a stream for events related to patient registration that are published from the patient service and consumed by the inspection service. Once we do that, each service can operate at its own speeds and come and go at any time. This is because the stream takes care of at least once or exactly once delivery guarantees. If we think about a scenario where our applications are running on a serverless platform within ephemeral containers that come and go arbitrarily, this kind of approach would help us not lose any messages that are published to the NATS server. There are several other capabilities available in JetStream, such as message replay and different message expiration policies, that we can utilize to build a reliable and efficient application based on the microservice architecture.

Since JetStream is an additional feature that can be enabled easily in the NATS server and supports clustering and horizontal scaling, we can create a flexible deployment with a set of servers acting as core NATS servers that provide maximum performance, all while another set of servers acting as JetStream-enabled servers take care of the persistence and message delivery guarantees. Since NATS allows messages to be published to any server and routed to the JetStream-enabled servers internally if there is a route, clients do not need to worry about publishing to specific servers.

Using the saga pattern to develop applications with transactional requirements

There can be many real-world use cases that require a sequence of operations to be executed in a transactional manner. The order processing scenario we discussed in the *Saga pattern and managing data* section is such an example. In *Chapter 9, How Microservices and NATS Coexist with Integration Platforms*, we discussed a solution architecture where microservices are integrated with an integration platform. We can utilize the integration component of that architecture to act as the orchestrator to implement the saga pattern. To do that, our microservices that are engaged in transactional operations need to improve so that each of those services has the compensating operations built into them. This concept is depicted here:

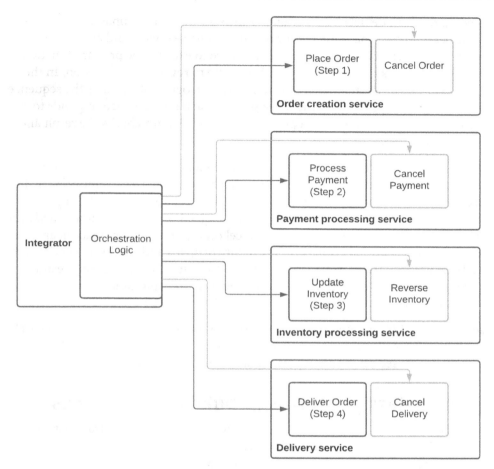

Figure 10.5 – The saga pattern implemented with the integration component

The preceding diagram depicts how we can use the saga pattern to implement a transaction-based order processing application using microservices and an integration component. Each microservice offers an API resource to execute the primary function of that service, as well as a compensating API resource to reverse the operation. In the integration component, we implement the orchestration logic that executes the sequence of operations (functions) by calling the microservice. Each microservice responds to the orchestrator with the result of the operation. The orchestrator checks the result and decides on the next step.

As an example, let's assume that a customer has placed an order to purchase a mobile phone and that the price of the phone is higher than the customer's account balance. In such a scenario, the order placement process is initiated and done before the payment processing operation. Hence, if the payment processing fails, the orchestrator will identify the failure, call the compensating operation (cancel order) of the order creation, and make sure that the order is canceled properly. If the failure happens at the delivery service level due to an issue with the travel restrictions or due to a technical failure, the orchestrator will call the compensating operations of all the previous three microservices and reverse the order in the system.

Likewise, we can build transactional applications with microservices using the approach and the architecture we have discussed within this book. It is up to you to decide on which approach to follow based on the application that needs to be designed and implemented.

Using serverless platforms to build microservices

One major challenge with building complex enterprise applications is infrastructure management. Most organizations have dedicated teams to manage infrastructure and these teams become a bottleneck when multiple teams need to release their applications to users. As we discussed in the early parts of this book, microservice teams are highly dynamic and operate independently. In such an environment, having a rigid infrastructure management process would jeopardize the entire effort and advantages of the microservice-based application's design. This is one of the main reasons why we need to look for a better approach to managing infrastructure. That is where serverless platforms can help with microservice-based application design and implementation.

Serverless platforms provide an additional level of comfort to manage infrastructure when building applications as microservices. The microservice teams can spend their time implementing the business logic of the microservice without worrying about infrastructure management. Once the microservice has been developed, it can be deployed as a function within the serverless platform. The serverless platform takes care of resource allocation in terms of memory and CPU, as well as the observability aspects of the service and the scalability aspects as and when required. In addition to that, serverless platforms typically run applications in an ephemeral manner. This will reduce the overall cloud infrastructure costs and help utilize the underlying computing resources more efficiently.

Hybrid deployments to support brownfield enterprises

In *Chapter 9, How Microservices and NATS Coexist with Integration Platforms*, we discussed that many other systems coexist with the microservice applications in a real-world enterprise platform. Even though cloud computing infrastructure offers many advantages for deploying applications, not every application and system can be deployed into the cloud. That is where the concept of the hybrid deployment architecture comes into the picture. Most of the enterprise platforms are in a transitioning phase where architects and developers are trying to migrate applications as much as possible to the cloud. This lets them get rid of the infrastructure management overhead and reap the other benefits offered by the cloud. This means that most of these organizations will be living in a hybrid state in terms of the deployment architecture. The term hybrid means deploying applications on cloud infrastructure, as well as within an on-premises infrastructure that is managed by the organization itself.

Moving applications to the cloud is not a task that we can finish in a brief period of time. It is a long-running process that requires proper planning and sometimes involves a lot of trial and error. Hence, it is a journey but not a destination. In the enterprise architecture that we proposed in *Chapter 9, How Microservices and NATS Coexist with Integration Platforms*, there were several components that we could move to the cloud, as well as deployed in a hybrid model. The following diagram depicts the deployment options for those components:

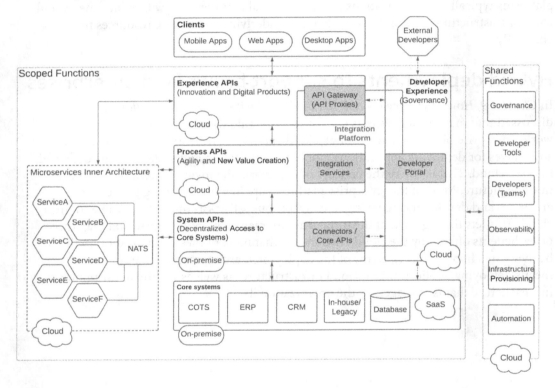

Figure 10.6 – Enterprise platform with microservices and integration deployment options

As depicted in the preceding diagram, there are certain components that we can migrate and deploy in a full cloud environment, such as an **Infrastructure as a Service (IaaS)** platform or a serverless environment. The following components can be deployed entirely or partially in such environments:

- **Microservices**: This is a component that can be deployed fully on the cloud. Some microservices such as stateless microservices can be deployed as functions in a serverless environment. Other microservices can be deployed as containers or VMs within the cloud infrastructure.

- **NATS server cluster**: We can deploy NATS servers in a cloud environment within a VM or containerized platform. It is not suitable to run NATS servers in a serverless platform due to the ephemeral nature of the serverless containers. If required, NATS servers can also run in an on-premises environment if you have enough resources to manage it. In addition to that, NATS can be used as a **Software as a Service (SaaS)** from a vendor.

- **Core systems**: Most of the time, these core systems run in an on-premises infrastructure within secured networks. These systems are not connected to the public internet and these systems can only be accessed through gateways and firewalls by external users. These systems have specific data protection and security requirements that prevent them from running in a public cloud environment. Some of these core systems are available as SaaS products and in such situations, you can use those options since the vendor takes care of data security and protection.

- **Integration platform**: This is a component that can be deployed in the cloud as a whole so that all the major components including the system, process, and experience APIs and the developer portal can be deployed in an IaaS platform. Another option is to use it through a SaaS offering provided by the vendor. Another common approach that's available with integration platform vendors is the ability to deploy certain components such as a developer portal and analytics in the cloud, as well as other runtime components such as the system, process, and experience APIs and security components on-premises so that the integration platform itself becomes a hybrid deployment.

- **Shared functions**: These shared functions can be deployed in the cloud within IaaS platforms. Most of the cloud service providers have functions such as automation, provisioning, observability, and developer tools built into their cloud offering so that these functionalities can be enabled when using their service to deploy applications. You should use the services offered by the vendors for these functions as much as possible since those tools and functions integrate natively with those platforms.

These approaches can be used to improve our proposed architecture so that we can build microservice-based enterprise platforms with NATS messaging. Now, let's summarize this chapter.

Summary

In this chapter, we discussed some of the challenges that we may face when designing and implementing microservice-based applications using NATS as the messaging layer. We looked at some of the ways that we could address those challenges and make sure that our microservice-based applications, as well as the overall enterprise platforms, are designed and implemented properly. Then, we went through some of the newer developments happening in the microservice architecture, where we looked at service meshes, the saga pattern, and serverless platforms. After that, we discussed some of the new features that have been introduced with the NATS platform recently. Specifically, we discussed the JetStream capability, which allows us to build applications that require higher message delivery guarantees than the core NATS server, which we discussed in the previous chapters.

We demonstrated the key capabilities of JetStream by trying out a few examples. Then, we discussed how we can improve our proposed solution architecture and approach to building microservice-based applications and platforms with NATS by utilizing some of these newer developments. There, we discussed using JetStream to avoid message losses and using the saga pattern to implement transactional applications. After that, we went through the hybrid deployment pattern that we could follow when building real-world enterprise platforms with microservices. We also took a brief look at how to utilize serverless technologies when building microservice-based applications.

This marks the end of this book, and we hope you learned something new by reading it. We learned how a microservice architecture is built, integrated, and managed within an enterprise information system. We started by looking at the concepts of microservices and NATS messaging, and then went on to build an architecture that could be used to build an enterprise information system with microservices.

Then, we demonstrated, along with code samples, how to build such a platform using a hypothetical **outpatient department** (**OPD**) application that is used in a hospital. We went from looking at the core microservice principles to advanced concepts such as security, observability, and integration. We went through a few examples along the way to display these concepts. Finally, we discussed some improvements and developments that would be useful in the future when designing enterprise applications and platforms with microservices.

Packt.com

Subscribe to our online digital library for full access to over 7,000 books and videos, as well as industry leading tools to help you plan your personal development and advance your career. For more information, please visit our website.

Why subscribe?

- Spend less time learning and more time coding with practical eBooks and Videos from over 4,000 industry professionals

- Improve your learning with Skill Plans built especially for you

- Get a free eBook or video every month

- Fully searchable for easy access to vital information

- Copy and paste, print, and bookmark content

Did you know that Packt offers eBook versions of every book published, with PDF and ePub files available? You can upgrade to the eBook version at packt.com and as a print book customer, you are entitled to a discount on the eBook copy. Get in touch with us at customercare@packtpub.com for more details.

At www.packt.com, you can also read a collection of free technical articles, sign up for a range of free newsletters, and receive exclusive discounts and offers on Packt books and eBooks.

Other Books You May Enjoy

If you enjoyed this book, you may be interested in these other books by Packt:

Cloud Native Applications with Ballerina

Dhanushka Madushan

ISBN: 9781800200630

- Understand the concepts and models in cloud native architecture Get to grips with the high-level concepts of building applications with the Ballerina language Use cloud native architectural design patterns to develop cloud native Ballerina applications Discover how to automate, maintain, and observe cloud native Ballerina applications Use a container to deploy and maintain a Ballerina application with Docker and Kubernetes Explore serverless architecture and use Microsoft Azure and the AWS platform to build serverless applications

Practical Cloud-Native Java Development with MicroProfile

Emily Jiang, Andrew McCright, John Alcorn, David Chan, Alasdair Nottingham

ISBN: 9781801078801

- Understand best practices for applying the 12-Factor methodology while building cloud-native applications
Create client-server architecture using MicroProfile Rest Client and JAX-RS
Configure your cloud-native application using MicroProfile Config
Secure your cloud-native application with MicroProfile JWT
Become well-versed with running your cloud-native applications in Open Liberty
Grasp MicroProfile Open Tracing and learn how to use Jaeger to view trace spans
Deploy Docker containers to Kubernetes and understand how to use ConfigMap and Secrets from Kubernetes

Packt is searching for authors like you

If you're interested in becoming an author for Packt, please visit `authors.packtpub.com` and apply today. We have worked with thousands of developers and tech professionals, just like you, to help them share their insight with the global tech community. You can make a general application, apply for a specific hot topic that we are recruiting an author for, or submit your own idea.

Share Your Thoughts

Now you've finished *Designing Microservices Platforms with NATS*, we'd love to hear your thoughts! Scan the QR code below to go straight to the Amazon review page for this book and share your feedback or leave a review on the site that you purchased it from.

https://packt.link/r/1-801-07221-3

Your review is important to us and the tech community and will help us make sure we're delivering excellent quality content.

Index

Symbols

A

www.ingramcontent.com/pod-product-compliance
Lightning Source LLC
Chambersburg PA
CBHW062054050326
40690CB00016B/3085